SUCH A LANDSCAPE!

SUCH A LANDSCAPE!

A Narrative of the 1864 California Geological Survey Exploration of Yosemite, Sequoia & Kings Canyon from the Diary, Field Notes, Letters & Reports of

William Henry Brewer

Introduction, Notes & Photographs by
William Alsup

Foreword by Cathleen Douglas Stone

Yosemite Association
Yosemite National Park, California

Frontispiece

THE SIERRA CREST AND MOUNT WHITNEY

Brewer and Hoffmann believed that Mt. Brewer was the highest peak in the range. On the other side, they had expected to see a downward slope to the Owens Valley. Instead, this view burst upon them, a sea of even sharper crags rising to an even higher crest, culminating in Mt. Whitney (far right). Brewer exclaimed: "Such a landscape!"

My ascent in 1981 was from the opposite side, from East Lake. I finally signed the register at mid-day and had the summit all to myself. I too marveled at the sea of sharp blades and deep gorges between me and the crest fifteen miles away. Even in the flat light of a cloudless noon, this image explains why Brewer and Hoffmann thought King and Cotter were mad to attempt to reach the crest.

Yosemite Association
Box 545
Yosemite National Park
California 95389

The Yosemite Association initiates and supports interpretive, educational, research, scientific, and environmental programs in Yosemite National Park, in cooperation with the National Park Service. Authorized by Congress, the Association provides services and direct financial support in order to promote park stewardship and enrich the visitor experience.

To learn more about our activities and other publications, or for information about membership, please write to the address above, or call (209) 379-2646.

LIBRARY OF CONGRESS CATALOGING-IN-PUBLICATION DATA

Brewer, William Henry, 1828-1910.
Such a landscape! : a narrative of the 1864 California Geological Survey exploration of Yosemite, Sequoia & Kings Canyon from the diary, field notes, letters & reports of William Henry Brewer / introduction, notes & photography by William Alsup ; foreword by Cathleen Douglas Stone.
p. cm.
Originally published: 1987.
Includes bibliographical references and index.
ISBN 0-939666-91-X (paper)
1. Sierra Nevada (Calif. and Nev.)—Discovery and exploration. 2. Sierra Nevada (Calif. and Nev.)—Description and travel. 3. Sierra Nevada (Calif. and Nev.)—Surveys 4. Yosemite National Park (Calif.)—Description and travel 5. Sequoia National Park (Calif.)—Description and travel 6. Kings Canyon National Park (Calif.)—Description and travel 7. Brewer, William Henry, 1828–1910—Journeys—Sierra Nevada (Calif. and Nev.) 8. Geological Survey of California. 9. Scientific expeditions—Sierra Nevada (Calif. and Nev.) I. Alsup, William H. II. Yosemite Association. III. Title.
F868.S5B74 1998
917.94'4044—dc21 98-27798
CIP

To the memory of William O. Douglas,
who would have liked Brewer

Contents

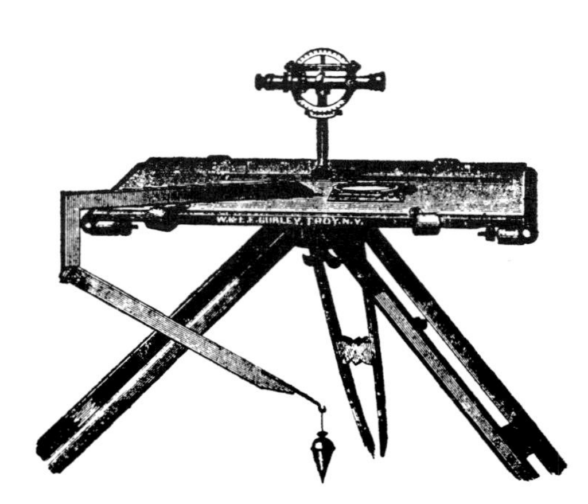

FOREWORD

William Alsup is a San Francisco lawyer, who over the past two dozen years has spent his spare time exploring the majestic Sierra Nevada of California. With view camera in tow, Bill has hiked much of the country covered by the 1864 exploration of the Sierra conducted by the California Geological Survey, whose field party was led by William H. Brewer.

The Survey, commissioned by the California Legislature, was charged to survey and map California and to report back a full scientific description of its soils, fossils, minerals, flora and fauna. Beginning on May 24 of 1864 and ending in September of the same year, William Brewer, Clarence King, Jim Gardiner, Charles Hoffmann and their packer Dick Cotter entered the then unknown and largely unmapped rugged peaks of the Sierra. They traveled without tents or weather-protective gear in temperatures ranging from below zero to over 100 degrees. Their journey was arduous.

Inspired by reading Brewer's spontaneous and factual account of the trip in *Up and Down California*, a volume containing Brewer's letters to his brother, edited by California historian Francis P. Farquhar in 1930, Bill began a series of treks to retrace the Survey's route. Bill tracked down Brewer's diary of the trip at the Bancroft Library at the University of California at Berkeley. There is a pocket volume of the diary for each year of Brewer's California survey work, which began in 1860 and ended in 1864. The library also has Brewer's field notebook and various letters by Brewer, in addition to those to his brother. Written in pencil, the soft script of the diary and field notes reveals moments of discovery and wonder—man discovering his world much as the Creator left it. Bill also digested the Survey's 1865 report, a remaining copy of which is located at the San Francisco library of the Sierra Club. Along the way, Bill made himself the Survey's photographer. That he was able to begin his task more than 100 years after the expedition speaks only to the timelessness of the Sierra. A mere blink separates Alsup and Brewer along their common route, especially when one considers that the rock under their feet was once covered by sea.

I met Bill when he was more of an indoor man. Specifically, he was working 12- to 16-hour days at the United States Supreme Court, where he was a law clerk to my late husband, William O. Douglas, for the 1971–72 term. Fresh from a joint-degree program of the Harvard Law School and Harvard's Kennedy School of Government, Bill was hired as law clerk to bring a broader perspective than that born of traditional legal training to the thorny issues that came before the Court. Now, Bill Douglas was a premier jurist, a boon companion and a friend of the Constitution, but his reputation as an employer has long been haunted by the suggestion that he was insensitive to anything other than the work needs of his law clerks. He remedied this annually by encouraging his law clerks to accompany him on the C&O Canal annual hike which is traditionally held on the last weekend in April. Hardly the Sierra Nevada, the C&O Canal is a stretch of tangled but flat wild from Washington, D.C. to Cumberland, Maryland, following the meander of the Potomac River.

The hike was at first an annual event to promote the creation of the C&O National Historic Park and, after it gained park status, to celebrate it. It is a two-day affair, beginning with a Friday-night cocktail hour, dinner, and an overnight campout. After a dew-drenched breakfast, the Saturday 10- to 16-mile hike begins. Some law clerks expressed a decided preference for the Friday-night activities. Bill Alsup went. It was his only day off for the entire year.

While at the Court, Bill worked with the Justice on his dissenting opinion in *Sierra Club* v. *Morton*. In that remarkable opinion, it was argued that inanimate objects, such as lakes and mountains, should, like other inanimate entities such as corporations, have standing in court, through a guardian *ad litem*, to argue their unique interests in lawsuits which will decide whether a stream should be dammed or a mountain leveled. Bill Alsup is responsible in great part for the footnotes in that dissent which the Justice insisted detail the role the Forest Service, which William Brewer helped to found, had come to play in the destruction of our national forests. Law and the land. Older man teaching, younger man reaching. A bond was formed, and it pleases me immensely that Alsup has dedicated this volume to Douglas.

Bill Douglas first encountered the Sierra Nevada range through the writings of John Muir, which he read as a boy. "It seems as if I have known the Sierra all my life," he once wrote. "I was with him [Muir] when the great winds blew and the pine and fir forests played their symphonies." Douglas later traveled throughout the Sierra, preferring the east-to-west approach over the great Sierra escarpment. He marveled at the Giant Sequoia and at the feisty, diminutive Douglas squirrel. He admired the resilience of the lodge pole pine, whose seeds survive fire and pestilence. He chronicled the high country wildflowers—the delicate penstemons, the cinquefoil and the bleeding heart. As I write, I can see Bill with his large, big-boned hands holding delicate petals under a tiny magnifying glass he customarily wore on a rope around his neck.

Thanks to Bill Alsup, we will know the Sierra anew through his lens and the terse but riveting words of William Brewer. By bringing both to us, more is accomplished than shared beauty. A tradition is being passed. Those who know a piece of the Earth come to love it. When modern life threatens to extinguish in mere decades what nature has taken millions of years to create, those who love the land will fight to protect it. Not everyone who fights will have climbed to the peak of Mt. Brewer, as Alsup did on a cool day in duly of 1981. But because of Brewer and Alsup, the Sierra will find new advocates.

CATHLEEN DOUGLAS STONE

Boston, Massachusetts

1 CALIFORNIA OAK, DIABLO RANGE

Leaving San Francisco on May 24, the Survey crossed to Oakland, mounted their horses, traveled south, crested the Diablo Range at Pacheco Pass on the 28th, and rode across the Central Valley on the Butterfield Stage Road, arriving at Visalia on June 4. In these twelve days, they witnessed one of the most severe droughts in California history. Brewer wrote repeatedly of intense hardship, parched carcasses, towering whirlwinds, and the blessed relief of shade under the oaks. The oak in this photograph is from the Diablo Range.

THE FIELD PARTY OF 1864

Gardiner, Cotter, Brewer, King

(Courtesy of the Bancroft Library)

INTRODUCTION

Such a landscape!

Saturday, July 2, 1864. At daybreak, Brewer and Hoffmann were up, gathering their gear. It was dark and cold. The tumbling stream was iced and the frosted grass slick beneath their boots. Camp 171 was at 9750 feet; it nestled in the dawn shadows of the peak that for days had dominated the horizon and which this day they meant to scale. They left. King, Cotter and Gardiner remaining, watched the pair fade eastward up the canyon.

A hundred peaks in sight
over thirteen thousand feet.

The peak rose like a bullet at the head of the canyon. It bristled with pinnacles. It yet had no name or place on any map. The pair believed the peak would prove to be the highest in the Sierra and the source of the headwaters for three mighty rivers.

Deep canyons, cliffs in every direction
almost inaccessible to man, on which
human foot has never trod.

Their first two attempts failed. The third succeeded. They were at once surprised and exhilarated—surprised because they had expected to stand atop the apex of the Sierra and to see an eastward descent to the great Owens Valley—exhilarated because they were far higher than they had anticipated. Instead of the Owens Valley, they beheld wave upon wave of cliff gorge rising eastward to even greater elevations, indeed, to the top of the nation.

Here is the highest and grandest group
of the Sierra—in fact, the grandest
in the United States.

It was such a group. And, it was an historic moment—the first recorded glimpse from deep in the Sierra of its massive crown—a moment shared that evening with their three companions below, a moment shared many times over through accounts of the ascent.

I

In the history of mountaineering in the Sierra Nevada, the 1864 expedition of the California Geological Survey deserves its special prominence. It was the first systematic exploration of the loftiest reaches of the Sierra—reaches dominated by Kings Canyon, Sequoia and Yosemite National Parks, but then only vacant space on the map. The Survey began to fill in the map, discovering and naming Mounts Silliman, Brewer, Whitney, King, Gardiner, Hoffmann, Gabb, Abbot, Williamson, Goddard, Tyndall, and the Palisades. The Survey also marked a conceptual shift in Sierra exploration. Whereas previous explorers had concentrated on finding routes through the Sierra, the Survey sought to study and to appreciate the high country for its own sake.

This does not, however, fully account for the affection that yet clusters to the name of William Brewer and the California Geological Survey. Countless Sierra evenings have been brightened by stories of the 1864 expedition—the discovery of the Big Trees, the first ascent of Mount Brewer, the daring thrust for Mt. Whitney, the long march toward Mt. Goddard, the glimpses of the vanishing Sierra tribes, and the historic outing with Frederick Law Olmsted. It was a happy coincidence that these survey members were the first. They set extraordinarily demanding standards in mountaineering—judgment, courage, selflessness, loyalty, industry, good humor, and respect for the wilderness. And, they have inspired us to sustain them.

II

In the decade after the discovery of gold in California, individual prospecting gave way to corporate mining, and the easy strikes gave way to painstaking exploration. At the same time, silver and quicksilver finds fanned the popular belief that California was rich with ore of all types. This, and a youthful faith in science, fostered the expectation that scientific exploration and surveying could not only produce much needed maps for a growing state but would pinpoint California's treasures.

Consequently, in April 1860 the California Legislature established the California Geological Survey. The statute named Josiah D. Whitney, a Harvard scientist, as the State Geologist. The statute directed Whitney

> with the aid of such assistants as he may appoint, to make an accurate and complete Geological Survey of the State, and to furnish, in his Report of the same, proper maps and diagrams thereof, with a full and scientific description of its rocks, fossils, soils, and minerals, and of its botanical and zoological productions, together with specimens of the same.

Whitney accepted the post. As his "first-assistant," he chose William H. Brewer.

III

William Brewer was born in Poughkeepsie, New York, on September 14, 1828. Reared on a farm, he entered Yale College in 1848 and studied agricultural chemistry under Professor Benjamin Silliman, receiving a degree in 1852 and graduating with the first class of the Sheffield Scientific School. Brewer then applied for a place on the Gunnison Expedition of 1853–54. Fortunately, his application was rejected, for as it turned out, the ensuing expedition was massacred by Indians. Brewer then taught in New York, studied and hiked in Europe, and returned to America to accept a professorship of chemistry at Washington College in Pennsylvania. He married, but in 1860, suffered the loss of his wife and infant son.

At this point, Brewer received Whitney's offer to be his first assistant on the California Geological Survey. Although the two had never met—the offer having been made on the strength of the recommendation of Yale Professor George Brush—Brewer desired a change. He accepted, and within a few months sailed with Whitney to San Francisco.

They reached San Francisco by steamer on November 14, 1860. Brewer wrote his brother that it was "by far the most beautiful harbor I have ever seen, not even excepting New York" and that San Francisco itself, although only ten years old as anything more than an outpost, boasted "large streets," "magnificent buildings of brick and granite" and "seemed at least a half century old." Brewer's first evening in the new city was spent observing the fireworks and processions in honor of the election of Abraham Lincoln, the news of which arrived in San Francisco by "Pony" on the same day Brewer end Whitney arrived. Brewer was resolutely pro-Union. During his years in California, while headquartered in San Francisco at Montgomery Street, he developed great admiration for Thomas Starr King, a Boston minister and Union orator called to the new Unitarian Church in San Francisco.

Within a few days, Brewer and Whitney had gathered various instruments and crew and sailed for San Pedro harbor. Upon arriving, they traveled inland to establish Camp 1 on a hill overlooking Los Angeles, "a city of some 3500 or 4000 inhabitants, nearly a century old, a regular Spanish-Mexican town, built by the old padres" on "a great plain." Their vista from the "hill over the town," was "one of the loveliest" Brewer had seen.

Camp 1 was the first of hundreds of encampments of the Survey while recording measurements and observations throughout California. In the Survey's first four years, Brewer led the field party, the more senior Whitney joining in only some expeditions. During his tenure, Brewer traveled over 15,000 miles in California through virtually every corner of the state, including the Coastal Range, Mount Shasta, boomtowns, the great San Joaquin Valley, and finally the Sierra. The field survey contingent consisted of as few as three and as many as six, including Brewer, with a membership that changed from time to time. Their names—Gabb, Hoffmann, King, Cotter, Gardiner, Whitney and Brewer—today grace peaks throughout the Sierra.

IV

It was not until 1863 that the Survey challenged the Sierra. Brewer and William Gabb, the Survey's paleontologist, left San Francisco on April 1. Brewer wrote his brother that they traveled by mule south through the "great San Joaquin plain" as "interminable as the ocean." After five weeks, they arrived in Fort Tejon, situated in a "pretty grassy valley, at an altitude of about 3000 feet above the sea, covered with scattered oaks of venerable age and size and of great beauty, with mountains rising around to the height of perhaps seven thousand feet, the valleys and hills green—picturesque, beautiful, quiet." Despite warnings of "Indian troubles," they explored the Kern River, Tehachapi Valley and Walker's Pass. Returning to the western foothills, they moved north via Millerton, to Hornitos, near Mariposa. Gabb then returned to San Francisco to write a report. Brewer proceeded to Murphys near the Calaveras Big Trees to join Whitney and Charles Hoffmann, a thin, pipe-smoking German who helped pioneer topography in America, principally through his work as the Survey's topographer.

Murphys was the staging point for the field party's first trans-Sierra expedition and its first exploration of Yosemite. In 1863, Yosemite was still a fresh discovery, having been entered only a dozen years earlier by the Mariposa Battalion. There was then no road and only the Coulterville and Mariposa Trails entered the Valley. Whitney, Brewer, Hoffmann and their packer, remembered by Brewer only as John, set out for Yosemite Valley on the Coulterville Trail via Big Oak Flat, Bower Cave, Deer Flat, and Crane Flat.

After reaching the Valley, Brewer wrote his brother that they camped beneath El Capitan, "the enormous precipice called Tu-tuc-a-nu-la," and later directly in front of Yosemite Falls. Brewer explored the Valley thoroughly, admired Mirror Lake, and climbed to the tops of both Nevada Falls and Yosemite Falls. His note that "there are only two or three parties here, perhaps a dozen persons in all," today staggers the imagination. How had the Valley been formed? The Survey pondered the issue, and Whitney advanced the thesis, which he defended the remainder of his life against John Muir and others, that the massive cliffs had not been caused by glacial action but by a cataclysmic event, or as Whitney put it, "the wreck of matter and the crush of worlds."

Leaving the Valley, the Survey retraced its steps along the Coulterville Trail to the Valley rim, there joining the Mono Trail by which miners had been reaching the Esmeralda District near the Nevada border. Near Porcupine Flat, the group climbed and named Mount Hoffmann in honor of their topographer. Passing Tenaya Lake, they camped at Soda Springs in Tuolumne Meadows. Brewer and Hoffmann named and climbed Mount Dana and named and challenged Mount Lyell,the highest point in the region that eventually became Yosemite National Park. Dana and Lyell were the "most eminent" American and English geologists.

All others having left for Murphy's, Brewer and Hoffmann observed the Fourth of July "by riding down the [Tuolumne] river a few miles and climbing a smooth granite dome for bearings." The view of the Cathedral Range and Tuolumne Meadows from this outlook, he thought, "would be a grand one for a painter—for a foreground we have a series of smooth, low granite domes, with grassy flats by them, with a cascade of the Tuolumne in front, a dark pine forest beyond stretching up against the slopes, while beyond and above are sharp granite pinnacles destitute of trees and streaked with snow." "All my adjectives are exhausted," Brewer wrote.

Brewer and Hoffmann continued through Mono Pass and Bloody Canyon to Mono Lake, inspected the boomtown of Aurora, and re-crossed the Sierra to the west via Sonora Pass. Hoffmann returned to San Francisco. Brewer joined Whitney at Calaveras for still another trans-Sierra crossing over Carson Pass to Lake Tahoe and more mining towns, returning via the head of Squaw Valley. Pleased with his two assistants, Whitney wrote that he "could not in the whole United States (or Confederate States either) find two men who would answer my purpose as well as Hoffmann and Brewer."

On Brewer's return to San Francisco by steamer on the Sacramento River, a serendipitous meeting occurred. Brewer wrote his brother: "On the way down two young men came up to me, asked if my name was Brewer and introduced themselves as two young fellows just graduated last year in the Scientific School at Yale College, who this summer have crossed the plains. Their names are Gardiner and King." Short, stocky, energetic and talkative, Clarence King had been drawn west by a reading of Brewer's letter describing his ascent of Mt. Shasta. King signed on as an assistant geologist. On January 22, 1864, Brewer wrote to Professor Brush at Yale that King "is a good fellow, tough, but doesn't like to do unpleasant work—yet I like him." James Gardiner enlisted as an assistant topographer and thereby joined King and the rest of the field party for the memorable expedition of the summer of 1864.

V

Brewer and the field party left Visalia on June 8, 1864, and "struck east." Their destination was the most rugged and least explored country in the Sierra. There they expected to find the top of the range and the sources of the three major rivers that poured onto the southern San Joaquin plain. The group consisted of Brewer, Hoffmann, King, Gardiner, and a packer, Dick Cotter. With them, King later wrote in *Mountaineering in the Sierra Nevada*, were "Nell and Jim, two old geological mules, branded with Mexican hieroglyphics from head to tail."

The campaign was inspired by a wintry glimpse of the high Sierra in January 1864. While surveying near Mariposa, Hoffmann climbed Mt. Bullion. On January 26, he wrote Whitney that although little could be seen of Yosemite Valley from that point, "there is another immensely high mountain mass south of [Yosemite] somewhat near the Owens Lake." Whitney believed that this might be the highest region in the Sierra. "This fact," wrote Brewer in the Survey's 1865 report, *Geology*, "coupled with the circumstance that, unless explored during the season by the Geological Survey, this region might long remain a blank on the map of California" compelled exploration in the summer of 1864 of the highest of the high Sierra.

To gather data to fill in the map, the party was equipped with barometers, compasses, levels and transits. The barometers measured elevations and were long glass tubes containing mercury whose height responded to the air pressure and hence the elevation. Before leaving Visalia, the Survey erected a reference barometer at nearby Camp Babbitt, at a known elevation, and measurements were taken by the Army through the summer. These measurements were later compared to the Survey's readings from the Sierra. Although the measurements tended to be 100 to 300 feet higher than modern measurements, they were remarkably accurate, considering the equipment. The level was used to determine which points within view were higher (or lower) than the level. The transit was used to establish a system of triangles to determine the relative location of landmarks. The topographical method developed by Whitney and Hoffmann (and adapted to general use for decades thereafter) established a series of precise triangles plotted from prominent peaks. The terrain profiles in between were sketched in by eye.

Leaving Visalia, the Survey ascended the foothills to Thomas' Mill at Mill Flat Meadow (since flooded for Sequoia Lake) and established Camp 164. This served as a base camp for a number of short explorations. One was to what is now called the General Grant Grove of Big Trees, discovered by Joseph Hardin Thomas in 1862. Brewer was exuberant in his description of the largest tree, later named the General Grant.

During short explorations from Camp 164, the group scouted the topography and developed a plan. Thomas' Mill and the giant sequoias, it was observed, stood on a broad divide between the watersheds of the Kings River to the north and the Kaweah River to the south, each river descending westerly to the San Joaquin plain. North of the divide was the precipitous drop into the great Kings River gorge with its red and gray walls. South of the divide was the gradual descent to the Kaweah River canyon. The divide itself continued to the east, gradually ascending, through a series of domes, meadows and forests. Thomas' Mill marked the last outpost of civilization. The way east was unmarked wilderness.

In the east, one distant peak stood out on what is now called the Great Western Divide. "Conspicuous upon the horizon, about due east of us," King later wrote in *Mountaineering*, was a "tall pyramidal mass of granite, trimmed with buttresses which radiated down from its crest, each one ornamented with fantastic spires of rock. . . . Upon the northern side it fell off, grandly precipitous, into the deep upper canon of King's River."

Although weeks away from it, the Survey party resolved to reach the tall pyramidal mass. It was, they thought, the source of the headwaters for the Kings, Kaweah, and Kern rivers and the apex of the Sierra. This peak ultimately was named Mount Brewer.

To reach it, they planned to stay high on the divide between the Kings and Kaweah rivers. For "experience had taught us," said King, "that the canyons are impassable by animals for any great distance; so the plan of campaign was to find a way up over the rocky crest of the spur as far as mules could go."

Following this plan, they toiled eastward along the ridge to Big Meadows. Soon, King recalled, they met two deer hunters who believed that "not even Indians had crossed the Sierras to the east, and that if we did succeed in reaching this summit we would certainly be the first." Pushing on, the party skirted Shell Mountain to the north, passed through Poop-Out Pass and followed the crest of the divide to JO Pass, where it established Camp 168 on Saturday, June 25. (These landmarks had no recorded names in 1864 and are referred to here by their modern names.)

The next day, Brewer discovered a major change in the topography. He discovered that the divide turned to the southeast and that the way to the east abruptly ended at a cliff. Brewer wrote in his diary: "After a late dinner I went up on the hill and came suddenly on a great granite amphitheater, 1600 feet deep with steep granite slopes or precipices on three sides and two pretty little lakes in the bottom [one Seville Lake]. The snowy peaks lay beyond partly obscured by clouds, a sublime view over the wild desolate landscape. Returned, got Gardiner and we went back and remained until darkness and clouds shut out the scene." This was the magnificent view of the Roaring River Basin and the Great Western Divide.

Imposed on its own divide a few miles to the southeast was a massive ruin that they named in honor of Professor Benjamin Silliman, Jr. On June 28, the group climbed Mt. Silliman. From this vantage point, Mount Brewer still loomed as the highest point on the horizon. The yet higher crest of the Sierra to the east was hidden, but barely, behind Mount Brewer and the Great Western Divide.

In the vicinity of Camp 168, the high ridge turned to the southeast and became increasingly steep and impassable. The party saw that the route had to be changed. Mt. Brewer was almost due east. It stood across the vast deep bowl of the Roaring River Basin, bounded by the Great Western Divide and the Silliman divide, each curving southward and meeting at Triple Divide Peak. The only feasible route was to abandon the high ground, descend into the basin, cross it, and ascend the canyon of the target peak as far as the animals could go. On June 30, the party moved into the basin and camped by Sugarloaf Rock.

On July 1, the group forded Roaring River and crossed Moraine Ridge. Having traversed the great basin, they pressed up the canyon of Brewer Creek. Camp 171 was established, according to Brewer's letter to his brother, about five miles west of Mt. Brewer at an elevation of approximately 9750 feet "by a rushing stream" with a "beautiful little lake" nearby and "poor feed" for their stock. Both Brewer and King noted that from this camp the conical peak stood tall against the eastern sky.

Camp 171 and Saturday, July 2, 1864, proved historic. "We were," Brewer wrote, "up at dawn." Brewer and Hoffmann set out to climb what they supposed was the highest peak in the Sierra. After two failures, they stood on the summit at two o'clock. A complete surprise burst upon them. Instead of a descent to the Owens Valley, as they had expected to see to the east, they glimpsed on the other side an expanse of titanic crags and deep gorges rising eastward, as King wrote in *Mountaineering*, to "a vast wall of mountains, lifting still higher."

"The view was yet wilder," Brewer exclaimed, "than we had ever seen before. . . . Such a landscape! A hundred peaks in sight over thirteen thousand feet—many very sharp—deep canyons, cliffs in every direction almost rivaling Yosemite, sharp ridges almost inaccessible to man, on which human foot has never trod—all combined to produce a view the sublimity of which is rarely equalled, one which few are privileged to behold."

Reaching camp at dusk, the party proclaimed the discovery and, no doubt, the camp blazed with excitement. "On calculating the peak, finding it so much higher than we expected, and knowing there were still higher peaks back, we were, of course, excited." This was "the highest and grandest group in the Sierra—in fact, the grandest group in the United States."

The next day, King begged Brewer to let King and Cotter try to reach "the top of California," as King put it. Brewer was reluctant because of the danger, pronouncing the plan madness. Hoffmann warned, according to King, that they "might as well attempt to get on a cloud as to try the peak." That night, however, Brewer confided to his diary that "King insists on footing it and I have consented against my judgment to let him and Dick try it—they are preparing to start in the morning. Long and animated has been the discussion and the matter has been discussed in all its aspects." In *Mountaineering*, King recalled his own doubts that night: "Brewer and Hoffmann were old climbers, and their verdict of impossible oppressed me as I lay awake thinking of it. . . . Never a man welcomed those first gray streaks in the east gladder than I did. . . ."

At dawn on July 4, they were off. Brewer and Gardiner assisted King and Cotter by carrying their gear to the shoulder of Mt. Brewer. King described the desolate sight from that point: "Rising on the other side, cliff above cliff, precipice piled upon precipice, rock over rock, up against sky, towered the most gigantic mountain wall in America, culminating in a noble pile of Gothic finished granite and enamel-like snow." When Brewer asked King what his plan was, "I had to own that I had but one, which was to reach the highest peak in the range." Away went King and Cotter. Brewer and Gardiner continued to the summit of Mount Brewer and planted the flag in honor of the day.

After four days of anxiety over King and Cotter, Brewer wrote in his diary: "Just before night, the joyful shouts of Dick and King." They unfurled a breathless account. Cotter's boots had been ripped apart, his feet wrapped in a flour sack. They had slept in snow. They had struggled against vertical cliffs. They had reached a summit over 14,000 feet. King had rung his hammer atop the summit and, ultimately, christened it Mt. Tyndall, for the geologist. They had seen at least two higher peaks. The highest they ultimately named Mt. Whitney, although King's notes on the summit show that he first called it Mt. Grant, in honor, no doubt, of the very popular Union general. Brewer wrote: "They crossed canyons, and climbed tremendous precipices, where they had to let each other down with a rope. It was by far the greatest feat of strength and endurance that has yet been performed on the Survey."

King was obsessed with climbing Mt. Whitney. When shortly thereafter the group returned to Big Meadows to restock their provisions, Brewer accompanied King on to Visalia. With Brewer's consent, King continued on south and made another attempt to reach Mt. Whitney via the partially constructed Hockett Trail. King resigned only three to four hundred feet short of the summit of Mt. Whitney. King rejoined the field party at the end of the summer at Clark's Station at Wawona. (King eventually climbed Mt. Whitney in 1873.)

Brewer was proud of King's effort. When in 1881 certain legislators tried to erase Professor Whitney's name from the peak, Professor Brewer protested vigorously, invoking King's exploration and the right of the Survey as discoverers to name the mountain. Brewer argued: "So far as one man could do it, I named the highest peak of that region Mt. Whitney in 1864, or, if that belongs not to me personally, then the exploring party of which I had command named it at that time." Stating that in 1864 he had felt it was more important to complete the reconnaissance to the north than to climb Mt. Whitney, Brewer wrote that he nonetheless sent King "to get on the peak if he could, with the scanty means I could furnish him with. Although then young and comparatively inexperienced, he was intrepid and enthusiastic, and poor as his means were, he nearly succeeded." No one else, Brewer reasoned, had "ever, up to that date, carried instruments so high as King got on Mt. Whitney." The governor vetoed the legislation and Mt. Whitney remained Mt. Whitney.

When Brewer returned to Big Meadows from Visalia on July 16, the Survey headed northeast toward Kings Canyon. They soon met an adventurous group of prospectors who had traveled from the Owens Valley via Kearsarge Pass west across the Sierra. Three were returning and showed them an old Indian trail (near Sheep Creek) leading to the deep and vast chasm of the Kings Canyon.

After descending into the gorge, Brewer pronounced it "next to Yosemite . . . the grandest canyon I have ever seen." They enjoyed two luxurious camps in Kings Canyon. The first was near Sheep Creek in a meadow near what is now Cedar Grove. The second was near the head of the canyon "by a fine grassy meadow where the stream forked."

For several days, they explored side canyons searching for a way out to the north. Their goal was to reach Mt. Goddard, which had loomed black and tall on the northern horizon (before their descent into the canyon), and whose height would assist in maintaining the continuity of their measurements and triangulations. Brewer explored

Paradise Valley and admired the Mist Falls. Finally, they pushed out of the canyon on its north side via the steep Copper Creek. They rose quite high to the divide itself and had a commanding vista. Mounts Clarence King and Gardiner, a dozen miles to the east, were thus discovered and named. They retreated, however, when the route proved impassable for the pack animals.

Back in the main canyon, Brewer decided to march due east across the Sierra to the Owens Valley and then re-enter the mountains farther north. Following this plan, the Survey took a rugged route eastward along Bubbs Creek (named for John Bubbs, one of the prospectors). They climbed the Charlotte Creek drainage and, at Kearsarge Pass, plunged over the precipitous eastern escarpment into the sage of the Owens Valley.

In the Owens Valley, the group moved northwest, parallel to the Sierra wall, until they found Mono Pass (not to be confused with the Mono Pass at the head of Bloody Canyon farther north). They crossed the pass and camped just below it at 9940 feet. From there the group explored nearby peaks, Gardiner going north to the Red Slate Mountains and Brewer going south to a summit on what Theodore Solomons later called Fourth Recess, one of the four hanging granite valleys above Mono Creek. Peaks were named for William Gabb, the Survey's paleontologist, and Henry Abbot, a member of the Williamson party of the Pacific Railroad Survey in 1855. They then followed Mono Creek to a camp in the beautiful Vermilion Valley.

Moving southward, the party again mounted a try at the dark omnipresence at the center of their wide arc—Mt. Goddard—in an effort to integrate their measurements for the summer. Despite a Herculean effort, they failed, realizing toward the end of a long day on August 10th that they had misjudged the distance. Hoffmann developed a sore foot and Brewer gave out, somewhere above Goddard Canyon at approximately 11,000 feet. Cold fell and they spent a freezing night beside a burning stump under a canopy of brilliant stars. This bivouac was Camp 192, or "Cold Camp" in the official reports. The map prepared by Hoffmann and published in 1873 (the "Hoffmann Map") shows the trail ending in the vicinity of Hell For Sure Lake. Incredibly, Cotter toiled on and came within 300 feet of the prize. He hung the barometer as the official report states, just in time to read it before darkness. All returned safely to camp the next day.

After returning to Vermilion Valley, the group struck northwest, crossing the basin of the north fork of the San Joaquin. By this point, all were tired. In camp, Brewer could muster little strength to write, stating only that the weeks had become day after day of "rides over impassable ways, cold nights, clear skies, rocks, high summits, grand views, laborious days, and finally, short provisions—the same old story." Worse, Hoffmann became ill, complaining of a pain so bad that he had to be lifted on and off his horse.

On August 23, they reached Clark's Station at Wawona on the Mariposa Trail to Yosemite Valley. They were greeted by Galen Clark and others, including Frederick Law Olmsted, the manager of the Mariposa Estate, the architect of Central Park in New York City, and soon to be the new Chairman of the first Board of Commissioners for Yosemite Valley. (Less than two months earlier President Lincoln had signed a law ceding Yosemite Valley and the nearby Mariposa Grove of Big Trees to the State of California for a public reserve, the first step in the ultimate establishment of Yosemite National Park.)

While Hoffmann convalesced, Brewer and Olmsted toured Yosemite Valley, Tenaya Lake, Soda Springs, Tuolumne Meadows and Mono Pass where they climbed and named Mt. Gibbs, a companion to Mt. Dana. On returning via Little Yosemite Valley, they found Hoffmann to be still flat on his back. Brewer, Gardiner, Cotter, and King, who had rejoined them at Clark's Station, carried Hoffmann by litter to Mariposa, thence by carriage to Stockton and thence by steamer to San Francisco. (Happily, Hoffmann recovered and later led other Survey parties in the Sierra.)

Back in San Francisco, Brewer received notice of his election as professor at Yale. After four years, he turned the leadership of the Survey's field work over to others, although he volunteered his own time in completing the botanical analysis of the Survey. In concluding his journal in California, he observed, "I have counted up my traveling in the state. It amounts to: horseback 7,564 miles; on foot, 3,101 miles; public conveyance, 4,440 miles—total, 15,105. Surely a long trail!" Indeed.

VI

After the 1864 expedition, the Survey lost momentum and was discontinued in 1874. Whitney was not popular in Sacramento, and the Survey was constantly looking over its shoulder in fear of losing its annual appropriations. Legislators ultimately felt that the Survey, despite its actual scientific and topographical accomplishments, had not adequately fulfilled the principal goal of many legislators—to locate mineral reserves in California. When its sequel was eventually established, the Legislature even named it the Division of Mines to disassociate it from the Survey and to stress its mineral orientation.

Time, however, has solicitously polished the image of the Survey. In 1872, a series of articles by Clarence King on the 1864 campaign was collected and published as *Mountaineering in the Sierra Nevada.* King thrilled his readers. His imagination did not sleep in inventing images that danced across the granite heights. Generations have smiled with King and have eagerly forgiven him for embroidering to make a good story.

Although Brewer left it to King to popularize the story of the summer of 1864, Brewer in fact kept an extensive personal account throughout his four years with the Survey, including the summer of 1864. Brewer maintained a daily diary and carried a field notebook. As time permitted, he composed a long series of letters from the field to his brother, each numbered sequentially, and all subject to a request for preservation. The letters were edited by Francis P. Farquhar, the foremost Sierra historian, and published in 1930 as *Up and Down California.* The diary and field notes are in the Bancroft Library at Berkeley, as are typescripts of the letters. (The original letters are at Yale, as are certain other Brewer papers mentioned here.)

As the following pages show, Brewer's record, even if less audacious, competes mightily with King's account. It chronicles a remarkable experience of one who saw more of the Sierra than virtually anyone since. If Brewer's adjectives were less vivid than King's, they were spontaneous. Written in the field, as judgments and history were made, Brewer's words mirrored a drama in progress. On the first ascent of Mt. Brewer, an ecstatic Brewer scribbled his impressions in his field notes at an angle across the sheet. The night before King and Cotter walked into Sierra history, Brewer inscribed his reluctance in letting them go. When their "joyful shouts" reached the camp five nights later, Brewer's diary captured a sense of relief that still radiates from the page.

This short book of vintage words and fresh images commemorates William Brewer, the Survey and the spirit of their summer of 1864. This is done, I would like to think, by three small contributions to Sierra literature. First, the following text of the journal includes material hitherto unpublished (Brewer's diary, field notes, and other correspondence as well as certain original notes of Clarence King and others) and material last published in 1865 (Brewer's report in *Geology*). Second, the text combines and collates all Brewer's various accounts of the 1864 expedition, including his letters, into a single day-by-day narrative in Brewer's own words, citing, for the sake of historical accuracy, the specific source of his writings. In addition to Brewer's writings, the records pertaining to the same day left by Hoffmann, King, Gardiner, Whitney and Olmsted are excerpted and referenced in italicized notes immediately following Brewer's narrative for the same day. Some of these materials, such as King's field notes (preserved at the Huntington Library), have not been published before. This day-by-day collation offers as accurate a chronological re-creation of the footsteps of the Survey as all available records will permit. Finally, I have had the good fortune to hike most of the original 1864 route, comparing my notes with theirs, and sharing the same places and scenes that excited them. Their original route and observations inspired the photographs that follow. I hope they capture the mood and character of that timeless landscape and, in turn, commemorate the mood and character of those five explorers who inspired a standard of mountaineering as demanding and as durable as the Sierra itself.

WILLIAM ALSUP

Oakland, California

All of the following journal entries are William Brewer's words. The specific source is indicated by a citation to a D, L, or R in brackets, indicating that the text preceding it is from Brewer's field diary, from his letters (to his brother) or from his report (published in Geology *in 1865), as the case may be. Although drawn from different sources, each journal entry describes the specific day indicated.*

A few grammatical changes Brewer might have preferred have been made. Gardiner actually spelled his name "Gardner" at the time of the expedition and Brewer and others so spelled it. Gardiner, however, is used herein for consistency. Bracketed information has been added for clarification but parenthetical material is Brewer's. Some duplicative text has been omitted and some paragraphs merged. In no case has any meaning been changed.

The italicized notes following certain daily entries provide accounts or map references by King, Hoffmann, Gardiner, Olmsted or others for the same day. References to the Hoffmann Map are to Charles Hoffmann's first map of the region published in 1873. Place names in the italicized notes and the bracketed material are the modern place names. A complete list of sources appears at the end.

William Brewer's NARRATIVE

The field party of the California Geological Survey left San Francisco on May 24, 1864. The field party consisted of William Brewer, Charles Hoffmann, Clarence King, James Gardiner, and Richard Cotter. All were on horseback. Gear and provisions were carried by mule. They traveled light, without wagons, tents or other luxuries. For two weeks, during one of California's worst droughts, they crossed the great Central Valley. We join them on June 7, 1864, as they approach Visalia.

To the Big Trees

Tuesday, June 7

We came on to Visalia, twenty-four miles. In a few miles we passed the belt of oaks that skirts the river for a couple of miles on each side; then across the barren, treeless plain, still perfectly level, in places entirely bare, in others with some alkali grass. The surface of the soil was so alkaline that it was crisp under the horses' feet, as if covered with a thin sheet of frozen ground.

Before reaching Visalia we again struck timber. The region about Visalia is irrigated from the Kaweah River, and is covered with a growth of scattered oaks—fine, noble, old trees. The town is a small place on the plain, but very prettily situated among the fine trees. [L]

Maj. Williamson arrived from Fort Tejon—called on Capt. Roper of Camp Babbitt, to see about escort, if needed. [D]

In 1853 Secretary of War Jefferson Davis commissioned then Lt. R. S. Williamson of the Topographical Engineers to explore all passes leading from the southern Central Valley to the Los Angeles Basin or to the Great Basin, including Walker Pass and Tejon Pass, and to make recommendations concerning their suitability for a railroad. Williamson's Report of Exploration in California for Railroad Routes *in 1854 was, a decade later, the leading work on mountain exploration in California. A few weeks after their June 7th meeting with Major Williamson, the Survey placed his name on a prominent peak near the crest of the southern Sierra.*

Wednesday, June 8

We struck east until we reached the hills, then up a canyon northeast. On the plain we soon passed the belt of trees, then over the barren plain, so dry and so hot. The heat was so bad that it nearly made us sick. The low foothills are as dry as the plain—no grass, but they are covered with a growth of scrubby oaks and bushes scattered over them. [L]

At last camped where we found but little feed, and poor water [Camp 162]. We had a glorious night's rest. [D]

Thursday, June 9

We were up at dawn and off early, and at noon came to camp [163] by a stream of delicious water, where our hungry animals had good grass. [L]

After a lunch I climbed a mountain near which commands a fine view of the whole western slope of the Sierra, the outer foothills which we have passed with their irregular topography, covered with scattered oaks and patches of chapparal, then the belt of conifers in which we are, and behind the bare, desolate and very rugged snowy peaks of the crest. The view grand. The great Valley fades away like the sea. Hoffmann killed the first rattlesnake. [D]

FRIDAY, JUNE 10

We came on but four miles to this camp [164]. Up, up, up, over a high ridge, and at last into a dense forest of spruces, pines, firs and cedars. [L] The belt of pines is entered quite suddenly at an elevation of from 3200 to 3500 feet and at 5000 feet the forests are already made up of large trees. [R] We then sank into a little depression where there is a beautiful grassy meadow of perhaps 200 acres, surrounded by dense, dark forests, where there is a steam sawmill, and where two or three families live. [L]

Found a great relief in the cool, balmy mountain air. Night cold, thermometer sank to 23 degrees. Our camp is over 5000 feet in altitude. [D]

The route from Visalia to Camp 164 was, according to the Hoffmann Map, along what is today State Route 69 and what was then a wagon road. Thomas' Mill was in a meadow since flooded to form Sequoia Lake just southwest of and below the Grant Grove of giant sequoia trees.

SATURDAY, JUNE 11

Up early, morning very cold. Thermometer 23 degrees. Then went up on the ridge and saw some big trees. [D]

I will describe but two. The largest one standing is 106 feet in circumference at the ground and 276 feet high [The General Grant]. But it swells out at the base, so that at 12 feet from the ground it is only 75 feet in circumference. It is finely formed and you can but imperfectly imagine its majesty. It has been burned on one side, and were it entire, its circumference at the base would be 116 to 120 feet!

Now for the other tree. It is prostrate and no larger, but the story seems bigger. It has been burned out so that it is hollow, and we rode into it seventy-six feet and turned around easily. For 40 feet *three* horsemen could ride in abreast, but we had but one horse along, which we took up on purpose to take this wonderful ride. Nor is it difficult, for most of the cavity is nine feet high and as wide. The greatest width is 11 feet 6 inches, and the greatest height is 11 feet 8 inches in the clear. Our horse was very gentle, and in this part I stood erect in the saddle and could just fairly reach the top! The tree is broken in two places and fire has widened the fracture. At 120 feet from the base the tree is still 13 feet 2 inches in diameter inside the bark, and at 169 feet it is still 9 feet in diameter inside of the bark!

There are trees of this species of every size, many being over 12 feet in diameter. Two of the smaller ones have been cut and split into fence posts—how it takes away from the romance of them, using them for fence posts—and in a few years more many of the smaller ones were sawn into lumber by the mill here! [L]

About six miles east of Thomas' Mill, a rocky summit, called Bald Mountain, was ascended for the purpose of getting the first idea of the topography of the unknown region about to be visited. It was easy of access, although 7936 feet high, and offered a fine view of the neighboring country and the extended crest of the Sierra. Its position was at once seen to be on the great elevated divide between the waters of Kings River on the north, and the Kaweah on the south. This divide runs up to the snowy mountains at the summit of the chain, and appeared to terminate in the highest group of peaks, some twenty-five or thirty miles distant.

About eight or nine miles to the north, and several

thousand feet below, was the canyon of Kings River, which seemed precipitous and impassable. Some twenty miles to the northeast, this river divides into two branches, and the course of the northern of these is such that the observer on the summit of Bald Mountain can look directly into it. The view is most impressive. Granite walls with buttresses, pinnacles and domes rise perpendicularly from three to five thousand feet above the river, and above these the bare rocky slopes tower up, high above all vegetation, into regions of perpetual snow. Dark lines of trees wind up the ravines on the mountainsides, becoming thinner and more scattered, until they disappear altogether, the summits of the mountains rising far above all vegetation, barren and desolate. Such is the character of the divide between the main forks of the Kings River. [R]

Along the crest, 25 miles east, is the rugged snow-covered peak that we hope to explore. The western slope is rough in the extreme, and both the topography and aspect are unlike anything else I have seen. The region is so *very* rough that I am filled with anxiety as to the possibility of reaching it. [L]

Although the 1865 Report refers to "Bald Mountain" and the Hoffmann Map uses this name on a peak in the place indicated by Brewer, this place name has not survived. The point ascended was not, as one might think, what is today known as Big Baldy but was very likely a knoll yet unnamed rising 7646 feet, six miles east of Sequoia Lake. With respect to the trip to see the Big Trees, King's report to Whitney states that all went except King.

SUNDAY, JUNE 12

Thermometer 27 at minimum, rose to 60 degrees. A most lovely Sunday. Air cool and balmy. Sat under the old pines. Read some and wrote a letter.

King read a sermon. In front of our camp a pretty grassy flat with cattle and horses grazing and a clear stream running through it. A bold ridge beyond, clothed with pine and fir with here and there the grand form of some "Big Trees" looming above the forest.

The sick boy at the house was taken to town today in an oxwagon. [D]

King's field book notes that "Jim and I had a good long talk in the morning" and "in afternoon I read one of Robertson's sermons (The Greek)." Gardiner's correspondence with his family indicates that he was unusually devout, which may have contributed to Brewer's decision, as he noted on June 19th, not to work during Sundays on the 1864 expedition.

MONDAY, JUNE 13

A lovely clear night and a cool morning. 28 degrees. Day perfectly lovely, clear, warm, invigorating. In camp all day, lazily reading. Hoffmann and Gardiner went onto a mountain eight miles distant to get observations. Read the "Wandering Jew" most of the day. [D]

TUESDAY, JUNE 14

Another lovely day after a lovely night. King and Dick left early this morning to look out a way to the next camp. I stayed in and wrote and read, a quiet lazy day. [D]

The layover at Thomas' Mill was necessitated by the wait for provisions to arrive from Visalia by wagon.

WEDNESDAY, JUNE 15

As yesterday, in camp. Took latitude observations. King and Dick returned and report a practicable trail. [D]

King and Cotter left on the morning of the 14th. They rode northeast until they emerged onto a bare ridge where they climbed a cliff and made observations and sketches of Kings Canyon and the Great Western Divide to the east. In Mountaineering, *King described their first glimpses of the massive peak which dominated the eastern horizon for weeks to come and which the Survey ultimately named for William Brewer: "Conspicuous upon the horizon, about due east of us, was a tall pyramidal mass of granite, trimmed with buttresses which radiated down from its crest, each one ornamented with fantastic spires of rock. Between the buttresses lay stripes of snow, banding the pale granite peak from crown to base. Upon the north side it fell off, grandly precipitous, into the deep upper canyon of Kings River."At this point, the Survey believed that the divide and Mt. Brewer were the highest in the Sierra, although King later claimed that he had suspected yet higher peaks lay beyond. The Survey's first major goal was to reach and to climb Mt. Brewer. Their plan was to ascend eastward along the gentle divide between the Kings and Kaweah rivers to the Great Western Divide.*

THURSDAY, JUNE 16

Our provisions have come and [we] made preparations for departure tomorrow. [D]

To Mount Silliman

FRIDAY, JUNE 17

We left all trails behind at the mill, but we had looked out a way beforehand. We struck back on the divide between the Kaweah and Kings Rivers, where an old Indian trail formerly ran. We divided our baggage between all the animals and walked, for the way was terrible. At times it lay over and along the ridge, in forests of fir and pines, and then over rocky hills and up steep slopes—so steep that our animals could hardly cling to them. We passed [and left behind] hundreds of the "Big Trees,"which are everywhere scattered through the forests here.

At last we struck a little meadow surrounded by forests, where we camped [at Camp 165 near Log Corral Meadow, placed by King's report at the junction of the Bald Mt. spur and the main divide]. It was at an altitude of about 7,400 feet. I went beyond, on a granite knob, where I had a grand view over this rough region, with the snowy peaks

2 BIG TREES, GRANT GROVE

The Survey's first significant camp was No. 164 at Thomas' Mill. This happened to be just below one of the most remarkable stands of giant sequoias, the Grant Grove, discovered in 1862 by Joseph Hardin Thomas. The Survey explored the grove on June 11. The largest tree described by Brewer was the tree named (in 1867) to honor General Grant.

This pattern of elegant and slender trunks, mostly sequoia, lies just below the General Grant. The fallen monarch in the foreground is very likely the hollow log Brewer entered on horseback. It is so large that it served as a home for two brothers for several years in the 1870s. Their fireplace eventually burned the opening on the trunk shown in the photograph.

ahead, which gave me a lively sense of the difficulties we would have to surmount. [L]

SATURDAY, JUNE 18

We came on here [Camp 166], about eight miles farther, over a region fully as rough, sometimes through forests, and at others over and among rocks. We at last struck a trail that has recently been cut to bring in cattle. We came to camp here, by a little meadow, where our animals have good grass, and we plenty of wood and water. It is at an altitude of about 7,800 feet. Here is a succession of grassy meadows—one called the Big Meadows is several miles in extent—and some men have cut a trail in and have driven up a few hundred cattle that were starving on the plains. Back of these lie the sharp and snow-covered peaks of the crest. [L]

Brewer's diary adds that their arduous route that day passed through a "keyhole where the packs touched on both sides." The 1865 Report placed Camp 166 "about two miles below Big Meadows," probably at Rabbit Meadow. This region is characterized by stately fir and pine forests, broad meadows, gentle streams, and white granite domes once polished by glaciers.

SUNDAY, JUNE 19

We stayed in camp. I shall not work Sundays this year as I did last; the state can afford to do without it. I will not use myself up as I did last summer. [L]

A quiet day in camp. Read most of the day. Day lovely —it is cold, however. Took observations and King read a sermon. [D]

MONDAY, JUNE 20

Three of us [not including King, whose field notes say he went bear hunting] went on a peak about five miles east [probably Shell Mountain] to spy out the country. The view was grand—on the west the whole slope of the Sierra and the great plain, ending in haze—around us the roughest region imaginable—along in front the crest of

3 SIERRA CORN LILY

In describing the flora between the Big Trees and Silliman Divide, Brewer noted the dominance of fir and pine, observing "there is but little grass or undergrowth of either herbs or bushes." One plant, however, most elegant in the early summer, is the Sierra Corn Lily (or False Hellebore).

the Sierra, its more prominent points not less than 12,000 feet high, with rocks, precipices, pinnacles, canyons, and all the elements to make a sublime landscape. We were up about 9,700 feet and it was very cold—only three degrees above freezing—and yet there are trees three to four feet in diameter. We had a weary and rough walk back by missing the way; but a hearty supper awaited us. [L]

In this part of the mountains, as at the Yosemite, the granite exhibits a tendency to form dome-shaped masses on a grand scale; but on the very crest or summit range [approximately 20 miles eastward], it rises in pinnacles, giving a very different character to the scenery. [The dome climbed] was made up of concentric layers or beds of granite from one to five feet thick, having no trace of gneissoidal structure, but breaking into large rectangular masses, sufficiently smooth and regular in form to be used for masonry, without dressing. [R]

TUESDAY, JUNE 21

The longest day of the year, ought to have been one of the hottest. I stayed in camp alone. Dick and King went out to shoot a bear, if they could. Hoffmann and Gardiner went on a ridge for bearings. At about noon it began to snow violently, and it continued all the afternoon. You cannot imagine how uncomfortable it is without any shelter. [The Survey had no tent.] The boys got back wet and numb with the cold. I had made a big kettle of soup, which was pronounced an eminent success. [L]

WEDNESDAY, JUNE 22

A very heavy frost fell and we were not up early. [D]

Before we started on this trip I heard that there were hostile Indians somewhere in here, driven out of Owens Valley. So I wrote the Governor, and he to the Commandant of the Pacific, who in turn issued orders to the various military posts within 200 miles of here to furnish me with a military escort if I should demand it. At Visalia there is a company of cavalry. The order had been received and the soldiers were very anxious to get away and begged me

4 THE PRECIPICE AND THE SILLIMAN DIVIDE

Ten days east of the Big Trees, the Survey emerged from the domes and forest and discovered that they were at the edge of a deep and long precipice. On June 26 Brewer wrote in his diary: "After a late dinner I went up on the hill and came suddenly on a great granite amphitheater, 1600 feet deep with steep granite slopes or precipices. . . ." This was the Silliman Divide. The group climbed Silliman on June 28.

In 1984, I climbed Mt. Silliman and saw, as had the Survey, that the highest peak on the horizon did appear to be Mt. Brewer on the Great Western Divide, and that Mt. Whitney and the true crest were hidden from view. This photograph is the view of the precipice of the Silliman Divide and Mt. Silliman, bathed in clouds, from near Silliman Pass.

5 THE GREAT WESTERN DIVIDE

From their camps along the Silliman Divide, the Survey seemed all the surer that the Great Western Divide was the highest crest in the Sierra and that Mt. Brewer was the highest point in the Sierra. This view is of the Great Western Divide and the undulating ridges of the Roaring River Basin from Silliman Pass at dusk. The distant peaks from left to right are North Guard, Mt. Brewer, South Guard, Thunder Mountain, and Table Mountain.

to make requisition. I did not, hoping that I would not need them. This morning two men from Visalia were here, an underofficer and a private, who wanted to go. As it will please the men, cost no one anything, more than it would to have the soldiers stay in camp, and as I thought they might be useful even if we found no Indians, I sent in a requisition with one of them for an escort to join us in two weeks, when we will strike north into the region possibly hostile, but most probably not. [L]

On June 22, Brewer began a letter to Professor George J. Brush of Yale, to which he added two later entries before mailing. The entry for the 22nd states in part: "I have a little leisure time today so I will write you a letter, although my fingers are entirely too cold for comfortable writing." After describing the general route thus far, he added "I have a little party of four beside myself, small but plucky. Clarence King is with me and he is a capital fellow in the field, and, by the way, a good joke." The letter described how King had been won over from the metamorphic theory of formation of Sierra granite to the eruptive theory of formation. Brewer also described the region, stating that the range rose to "peaks perhaps 12,000 ft. high," indicating the extent of the surprise in store a few days later in the ascent of Mt. Brewer and the discovery of the even greater heights of Mt. Whitney. Brewer commented: "You have no idea of the rough, laborious life we lead, without tents or shelter, our baggage cut down to the smallest possible allowance, sleeping on the ground, although it freezes every night. The last five nights the thermometer has ranged from 16 to 20 degrees F. in a region as rough as Switzerland but without trails."

During the layover at Big Meadows, King returned briefly, according to his report to Whitney, to Thomas' Mill to have his horse, Kaweah, reshod.

THURSDAY, JUNE 23

Got ready to start but we wanted some fresh meat. [D] I went to a hunter's camp and got 40 pounds of dried venison and bear meat, and then went on a hunt with him.

He shot one deer and had left three others where he had shot them a few days before. He gave me the meat of two and packed them out where we could get them the next day.

On my way back to camp I got bewildered and lost—for the first time in the state. I walked about two miles to a granite ridge where I could see the country, saw where I was, and then went to camp. It could hardly be called "lost in the Sierra," as I was not detained an hour, yet it was uncomfortable for the time. [L]

Returned to camp but too late to start. [D]

FRIDAY, JUNE 24

We cached all the baggage that was not absolutely necessary, and came on about eight miles, over hills, through brush and forests, among rocks, and finally came to camp [167] in a little grassy meadow in a canyon. We were tired and the amount of venison we ate for dinner might seem fabulous were it stated. [L]

"After riding for several hours," King wrote of the same day to Whitney, "on a dry dusty plateau, where the sparse forest and dry bench grass offered us no inducement to camp, we at last came upon a little ravine, blessed with a cool brooklet, a good meadow of carisces and onions. Here, while Gardiner and Hoffmann went off for topography, and Brewer lay fondly over Bleak House, *Richard and I did bake many loaves of bread for the future." His field notes add that they "sang Sweet Betsy at night."*

SATURDAY, JUNE 25

We came on about eight miles farther [to Camp 168 near JO Pass], and so rough was the way that we found this distance a good day's work. Our route lay along the divide between the head branches of the Kings and Kaweah Rivers, over steep ridges, some of them nearly 10,000 feet high, and then along ridges covered with forests of subalpine pines and firs. There are two species of pine and one of fir. All grow to a rather large size, say four to five feet in diameter, but are not high. All are beautiful, the fir especially so, but there is a difference enough in the color of the foliage and habit of the trees to give picturesque effect to these forests, which are not dense. All have a very dark green foliage, in harmony with the rugged landscape they clothe. There is but little grass or undergrowth of either herbs or bushes.

The rocks are granite, very light colored, the soil light—gray granite sand. Here and there are granite knobs or domes, their sides covered with loose angular boulders, among which grow bushes, or here and there a tree. Sometimes there are great slopes of granite, almost destitute of soil, with only an occasional bush or tree that gets a rooting in some crevice. Behind all this rise the sharp peaks of the crest, bare and desolate, streaked with snow; and since the storms, often great banks of clouds curl around their summits.

The whole aspect of this region is peculiar. The impression is one of grandeur, but at the same time of desolation—the dark pines, the light granite, the sharp cones behind, the absence of all sounds except the sighing of the wind through the pines or the rippling of streams. There is an occasional bird heard, but for most of the time silence reigns. At night the wind dies down, the clouds disappear, if any have occurred during the day, and everything is still. During the night there is no sound. The sky is very clear and almost black; the stars scarcely twinkle, but shine with a calm, steady, silvery light from this black dome above. [L]

SUNDAY, JUNE 26

Very quiet in our pleasant camp [168]. Wrote letters and prepared a glorious venison soup for dinner. [D]

Our present camp is by a little meadow, at an altitude of about 9,500 feet. The barometer stands at less than 21½ inches, water boils at 193½ degrees. Yet when one is still and not climbing he does not perceive the lightness of the air. It is a calm Sunday. The sky is intensely blue, a few white clouds float above, but it is cold in the shade, only 43 to 44 degrees, and my fingers are cold enough. Yet perhaps it is well, if it were more comfortable writing, I would perhaps write too much.

We have a book of sermons in camp, and thus far we have had one read aloud each Sunday. [L]

After a late dinner I went up on the hill, and came suddenly on a granite amphitheater, 1600 feet deep with steep granite slopes or precipices on three sides and two pretty little lakes in the bottom [probably Seville Lake and an unnamed smaller lake]. The snowy peaks lay beyond partly obscured by clouds, a sublime view over the wild desolate landscape. Returned, got Gardiner and we went back and remained until darkness and clouds shut out the scene. [D]

On June 26, Brewer added a further entry to his letter to Professor Brush, stating that they were camped in view of the "snowy peaks of the crest." Actually, this was the Great Western Divide, not quite the highest crest of the range, as they were soon to learn.

From Thomas' Mill, the Survey's plan had been to ascend eastward the gradual divide between the Kings and Kaweah rivers all the way to the crest. As Brewer's note on the evening of the 26th states, the way east abruptly ended at a cliff. The divide turned to the southeast, to Mt. Silliman. A vast basin lay between them and the "crest." A change in route was indicated. For the moment, however, the Survey continued to explore the Silliman Divide.

MONDAY, JUNE 27

Horses gone in the morning so Hoffman and I went on the peak to look out a way, while [the others] went after the missing animals. Soon finding them, we returned, packed up and came on to 169 but 3 miles and camped in a deep canyon by a little meadow. After dinner Gardiner and I went into the great amphitheater below while the rest went on the top. [D]

All the evening it rained. We turned into our damp blankets and prepared for a miserable night's sleep, but it stopped raining about nine o'clock and then cleared up, although the trees dripped water all night, and it was cold—the thermometer sank to 25 degrees. [L]

King reported that "by noon we reached a little meadow . . . and after supposing that we had made a fine march . . . were chagrined to find that we were not much out of long rifle range of our last camp." (The 1865 Report placed Camp 169 two, not three, miles farther along the divide.) Based on the Hoffmann Map, it seems clear that Camp 169 was between JO Pass and Silliman Pass in the meadow south southeast of what is today called Kettle Peak.

This interesting name was inspired by the kettle-shaped bowl surrounding Seville Lake, which Brewer called "the Kettle." King wrote Whitney that after dinner, he, Hoffmann and Cotter climbed north to the Kettle wall while Brewer and Gardiner climbed down a ravine to a lake in the bottom of the depression.

TUESDAY, JUNE 28

We had a fine clear morning, and four of us started to visit a peak a few miles distant. We had a rough trail, over sharp ridges, and finally up a very steep pile of granite rocks, perhaps a thousand feet high, to the peak, which is over eleven thousand feet high, and which we call Mount Silliman [11,188 ft.], in honor of Professor Silliman, Junior [the son of Brewer's Yale professor and a mining engineer in California].

In crossing a ridge we came on fresh bear tracks, and soon saw the animal himself, a fine black bear. We all shouted, and he went galloping away over the rocks and into a canyon. We had gone but a short distance farther when we saw a very large female grizzly with two cubs. She was enormous—would weigh as much as a small ox. After we looked at her a few minutes we all set up a shout. She rose on her hind legs, but did not see us, as we sat perfectly still. We continued to shout. She became frightened at the unseen noise, which echoed from the cliffs so that she could not tell where it came from, so she galloped away with the cubs. These would weigh perhaps 150 pounds each; she would weigh perhaps 900 pounds or more. We also saw a fine buck during the trip. [L]

This mountain is of granite, but of a darker color than the rock of the surrounding country; it is much intersected with veins of granite of a still deeper tinge than the main body of rock, and portions of it are reddened and discolored by the oxidation of the iron it contains. [R]

We reached the summit after a hard climb, and had a grand view of the rough landscape. Great rocky amphitheaters surrounded by rocky ridges, very sharp, their upper parts bare or streaked with snow, constituted a wild, rough, and desolate landscape. [L]

From the summit of this peak a magnificent view was obtained of the crest of the Sierra, as well as of the divide traversed by the party. The region to the east presented a complicated system of very sharp ridges, rising here and there into pinnacles, apparently all of granite, with numerous immense circular amphitheatral cavities, formed by sharp ridges surrounding basins, of which one side is always broken away, and which have exactly the appearance of ancient craters both in form and outline. To the west the predominance of rounded or dome-shaped mountain summits was most striking, the whole country having the appearance as if it had suddenly been cooled or congealed while violently boiling. [R]

Clouds suddenly came on, and a snowstorm, which was a heavy rain in camp. We got back tired enough [to a dinner, King wrote, made by Cotter, who had stayed behind]. [L]

This was the first major ascent of the expedition. From atop Mt. Silliman, the tallest peak on the horizon was Mt. Brewer on the Great Western Divide, standing across the vast basin from Mt. Silliman. This undoubtedly reaffirmed the belief of everyone on the Survey, with the possible exception of King, that the Great Western Divide was the culminating crest of the Sierra. Just hidden behind it, however, was an even taller crestline another fifteen

6 ROARING RIVER BASIN

To reach the Great Western Divide, it was necessary to descend into the Roaring River Basin, shown here from near the top of Mt. Silliman, looking north. The route was a circle round from the left to Sugarloaf Rock (shown in the upper left center) and then eastward (right). Mt. Goddard is the tall peak on the far horizon in the upper left. This photograph was taken during my climb of Mt. Silliman in 1984.

7 SUGARLOAF ROCK

One of the most prominent features of the Roaring River basin is Sugarloaf Rock. On June 30, 1864, the Survey camped beside it and climbed about it in the late afternoon. Brewer wrote that it "rises from the valley like a sugar-loaf." To explore the way, they climbed "nearly to the top," Brewer pronouncing its summit to be "quite inaccessible." However, he noted that there "is a magnificent view up the valley to the group of mountains forming the crest of the Sierra."

In 1986, my daughter, Allison, and I set out to cross the range, including the Great Western Divide, via some of the country covered by Clarence King and Dick Cotter in 1864. We entered at Sunset Meadow in Sequoia National Forest and exited at Onion Valley on the Owens Valley side. Our first camp was beside Sugarloaf Rock. In the late afternoon, I climbed as high as I was willing to go up this landmark. From this point, there was indeed a magnificent view of the Brewer Group. It almost certainly was the same place described by Brewer. From there, I made this exposure of the summit and glaciated sides of Sugarloaf Rock.

miles to the east, including Mt. Whitney, barely blocked from view by the trigonometry of an intervening, though shorter, barrier. The Survey, however, did have a fine view of the basin between the Silliman Divide and the Great Western Divide and could see the new route they would have to take to reach the other side. This route would take them north to Sugarloaf Rock and then directly east across Roaring River and up Brewer Creek Canyon.

King described the view in his field book and referred to names of peaks on the Great Western Divide as "Double Peak, Mt. Brewer's Peak, Table Mt. [and] the Crab's Claw." Farther north were "the Palisades and Mt. Goddard." Double Peak and the Crab's Claw did not survive as names. The rest appear on Hoffmann's 1873 map.

WEDNESDAY, JUNE 29

In camp [169] most of the day. King and Gardiner went to take a sketch of the higher Sierra. I went on ridge to sketch a parapet of granite by the head of the rocky amphitheater near with the fine view.

Our horses strolled off and went back to the last camp. Hoffmann and Gardiner went after them. Evening very mild. A change has taken place in the weather. It is warmer and the mosquitoes are out. Two bears and several deer seen by the party. [D]

Mount Brewer & Mount Whitney

THURSDAY, JUNE 30

We were up early and left. We changed our route and came on about ten miles, by such a terrible way that it was a hard day's work—over rocks, through canyons and brush. We sank into a canyon and camped [at 170] about two thousand feet below our last camp. [L]

The way to this camp led around the west and north sides of the Kettle [*i.e.*, around the rim of the bowl surrounding Seville Lake] over a region exceedingly difficult to traverse, with alternating steep, naked slopes of granite, and thick, low forest. [R]

A high rocky pinnacle by camp, a sharp granite cone [D], which rises from the valley like a sugar-loaf, as seen from below; but which, in reality, is the end of a ridge a mile or two in length. [R] Climbed nearly to the top of it to explore our way. [D] This is several hundred feet high and its summit is quite inaccessible. Its sides show undoubted evidence that it was once surrounded by a great glacier flowing down the valley. The slopes directed towards the moving ice are worn and polished, and huge boulders have been pushed up on them, and left all along, wherever the angle was not too steep for fragments of rock to lie. The meadow occupies a basin behind this knob, which appears to have been scooped out by a glacier. From the Sugar Loaf Rock there is a magnificent view up the valley to the group of mountains forming the crest of the Sierra. [R]

The Hoffmann Map shows the route passing south and west of the Kettle, starting from JO Pass to Rowell Meadow and thence to Sugarloaf Rock. King's report to Whitney stated: "Next morning we packed up and rode around the south side of the Kettle and across what appeared to be an old moraine, the animals picking their way among the old granite rocks with great skill. We then descended through an open forest . . . to the Kettle brook, passing through many beautiful, wood-surrounded meadows and reaching toward the middle of the day an elevated ridge [between the South and East Forks of Sugarloaf Creek] cutting across our way. We made a long detour around this and camped in a little meadow near the Kettle Brook [Sugarloaf Creek]."

FRIDAY, JULY 1

Up early and came on about 11 miles, but up a tremendous rocky hill over 1000 ft. high, then along the grandest moraine I have ever seen [Moraine Ridge]. In passing up it, the views of the canyon of the south fork of the river [Cloud and Deadman Canyons] very impressive, some of the cliffs almost rivalling Yosemite. Got into camp [171] tired enough. A picturesque camp but not a very

8 DEADMAN CANYON

On July 1, the Survey traveled from Sugarloaf Rock to Camp 171, the base camp for the ascent of Mt. Brewer. Brewer wrote in his diary: "Up early and came on about 11 miles, but up a tremendous rocky hill over 1000 feet high [Moraine Ridge]. In passing up it, the views of the canyon of the south fork of the river very impressive, some of the cliffs almost rivalling Yosemite."

In 1986, we too left Sugarloaf Rock and crossed the same river. (Brewer called the Roaring River the South Fork of the Kings.) Then we followed the Survey's general route up Moraine Ridge. At this point, we left all maintained trails behind. As we ascended the ridge into Brewer Creek Canyon, there were, as Brewer had written, several fine vistas up Deadman and Cloud Canyons. This view is of Deadman Canyon. We finally made it to a camp in Brewer Creek Canyon which, I believe after three visits to the canyon, was very close to Camp 171.

9 MT. BREWER

On July 2, 1864, Brewer and Hoffmann rose at dawn, followed Brewer Creek to its headwaters, and then ascended Mt. Brewer, finally succeeding on their third try. Our 1986 expedition followed the same route to the base of the mountain. (We did not climb the peak on this trip.) We decided to stop early and camped on a granite shelf at the base of the peak. This image was intended to record the way the mountain dominates the canyon.

good one. Feed scarce. Altitude 9750 ft. Back of us rise grand precipices with view of top of the high granite cone, bristling with pinnacles. [D]

Camp 171 earned its place in Sierra history as the base camp for the first ascent of Mt. Brewer and the discovery of the summit ridge of the Sierra. No longer was the Survey in domes and forests. This was timberline country. Just above them were talus fields, snowfields, the last grizzled pines, and then the rock and sky of the crest. Camp 171 was in one of the most desolate and scenic reaches of the range. Even today, there are no trails maintained in Brewer Creek Canyon.

Brewer, Hoffmann, and King left sufficient records to locate Camp 171. Both Brewer's diary entry for July 1 and letter of July 7 placed the camp at 9750 feet in view of the top of Mt. Brewer. (The Report placed it at 9969 feet.) The letter further placed the site "by a rushing stream" with a "beautiful little lake" nearby and five miles west of Mt. Brewer. The Report noted that Camp 171 was seven miles east-southeast of the previous camp at a point where "further progress with animals was entirely out of the question." Similarly, King's account noted that "by sunset we had gone as far as we could take the animals and in full view of our goal, camped for the night." The Hoffmann Map showed the route along Brewer Creek. Rockfalls and avalanches have taken their toll on the canyon but it seems clear that Camp 171 was along Brewer Creek and close by (and just southeast of) the small unnamed lake at approximately 9800 feet (which, in honor of the base camp, should be named Lake 171).

SATURDAY, JULY 2

We were up at dawn, and Hoffmann and I climbed [the] cone, which I had believed to be the highest of this part of the Sierra. We had a rough time, made two unsuccessful attempts to reach the summit, climbing up terribly steep rocks, and at last, after eight hours of very hard climbing, reached the top. The view was yet wilder than we have ever seen before. We were not on the highest peak, although we were a thousand feet higher than we anticipated any peaks were. We had not supposed there were any over 12,000 or 12,500 feet, while we were actually up over 13,600, and there were a dozen peaks in sight beyond as high or higher!

Such a landscape! A hundred peaks in sight over thirteen thousand feet—many very sharp—deep canyons, cliffs in every direction almost rivaling Yosemite, sharp ridges almost inaccessible to man, on which human foot has never trod—all combined to produce a view the sublimity of which is rarely equaled, one which few are privileged to behold. The view extended north 80-90 miles, south nearly as far. East we caught glimpses of the desert mountains east of Owens Valley, west to the Coast Range 130 miles or more distant. [L]

The summit is a loose and shattered mass of angular pieces of granite, forming a ridge some thirty feet long by five broad, which from the west appear as a sharp cone. The eastern side of the mountain is a precipice buttressed by a thin ridge, running east between two great vertically-walled basins, white with snow, which contrasts beautifully with the vivid blue of the frozen lakes 3000 feet below [including East Lake]. [R]

On our return we slid down a slope of snow perhaps eight hundred feet. We came down in two minutes the height that we had been over three hours in climbing. We got back very tired, but a cup of *good* tea and a fine venison soup restored us. [L]

Although Brewer's diary entry parallels his letter entry for July 2, his field notes, scribbled at the summit bear repeating: "Off at 6. Very clear. Coast range distinct over a great distance. [Illegible] haze. Last cone, very steep. Tried and failed. Up at last at 2. Top pinnacle of cone of granite—grand view. Snow group [Mt. Whitney Group]. Coast range. Slopes of the mts beyond Owens River in the ENE. All around the most desolate landscape. Scanty trees over 10,000 feet while several peaks are as high or higher than we are and there are hundreds in sight over 12,000. Whole [region] much more barren than the mts. N. The crest lies 10–12 m. [east] and there are 8–9 peaks from 100 to 400 to 500 feet higher than the granite cone we are on. In

10 EXFOLIATING GRANITE, MT. BREWER

In his report, Brewer described the laminations of Sierra granite, formed in the cooling process, and noted that the layers often break "into large rectangular masses, sufficiently smooth and regular in form to be used for masonry, without dressing." I happened upon this example in 1981 on the east ridge of Mt. Brewer, a few hundred feet below the summit.

11 MT. BREWER AND BREWER CREEK CANYON

"On July 4th, all were up at dawn. We got breakfast, and King and Dick packed their packs—six days provisions, blankets and instruments made packs of thirty-five or forty pounds each, to be packed into such a region!" Gardiner and Brewer helped carry the packs to give them a good start." We separated at the base of the cone, on the ridge, where they had a view of the terrible region they are to visit. Here we left them, and as we scaled the peak they disappeared over a steep granite ridge, the last seen of them." The point where the four divided was the ridge visible through the notch on the right shoulder of the peak. (There is another view of the notch and the ridge in Plate No. 5.) King and Cotter proceeded south (to the right). Brewer and Gardiner climbed north. Rating it as a Class 1–2 undertaking, the Climber's Guide to the High Sierra *traces the southwest route as follows: "From Brewer Creek climb to the notch just south of the peak and follow the ridge to the top." This was very likely Brewer's and Gardiner's route on July 4, 1864.*

the NE along the Mt. crest is a red shale that probably contains volcanic, all of the rest of the landscape is of light-grey granite. The most desolate region I have ever seen by far for several hundred square miles. But few trees, no meadows."

King recalled in Mountaineering *that "it was twilight and almost eight o'clock when they came back to camp, Brewer leading the way, Hoffmann following; and as they sat down by our fire without uttering a word, we read upon their faces terrible fatigue." As they were served supper, they told of yet another "vast wall of mountains" still farther to the east. "Hoffmann showed us on his sketch-book the profile of the new range, and I instantly recognized the peaks which I had seen from [atop Mt. Bullion near] Mariposa, whose great white pile had led me to believe them the highest points of California." King had been right—the Mt. Brewer wall was at the highest crest of the Sierra—still taller and wilder peaks—the Snow Group—were waiting.*

In a letter written later (January 11, 1873) to Professor Mayer in honor of Professor Tyndall, Brewer wrote: "That night [July 2] at the campfire (where our only tent was the deep blue-black vault of sky) the intrepid Clarence King earnestly begged to try with Cotter to reach the crest. I hesitated for we were short of provisions and far from supplies. Moreover, I had seen the difficulties and he had not, but he had read 'Tyndall's Glaciers of the Ages' and thought nothing was impossible."

SUNDAY, JULY 3

We lay until late. On calculating the height of the peak [placed at 13,886 feet, actually 13,570 feet], finding it so much higher than we expected, and knowing there were still higher peaks back, we were, of course, excited. Here there is the highest and grandest group of the Sierra—in fact, the grandest in the United States—not so high as Mount Shasta, but a great assemblage of high peaks. [L] We all became more anxious to know definitely the height of the higher peaks. [D]

King insists on footing it and I have consented, against my judgement to let him and Dick try it. They are preparing to start in the morning. Long and animated has been the discussion, and the matter has been discussed in all its aspects. Scarcity of provisions is a serious drawback but we will try and get over the difficulty. [D]

King is enthusiastic, is wonderfully *tough*, has the greatest endurance I have ever seen, and is withal very muscular. He is a most perfect specimen of health. [L]

The rest of the day was spent in preparation. Their gear and provisions included a barometer, a compass, a pocket-level, a set of wet and dry thermometers, rope, notebooks, bread, cooked beans and venison. These were rolled into blankets, which were twisted into two knapsacks with rope loops for their arms.

MONDAY, JULY 4

July 4th, all were up at dawn. We got breakfast, and King and Dick packed their packs—six days provisions, blankets and instruments made packs of thirty-five or forty pounds each, to be packed into such a region! Gardiner and I resolved to climb the cone again, as I had left instruments on the top, expecting someone would go up. Our way lay together for five miles, and up to thirteen thousand feet, I packed Dick's heavy pack to that point to give him a good start. I could never pack it as far as they hope to. [L] We separated at the base of the cone, on the ridge [elevation 12,800 ft.], where they had a view of the terrible region they are to visit. [D] Here we left them, and as we scaled the peak they disappeared over a steep granite ridge [South Guard], the last seen of them.

Gardiner and I reached the summit much easier than Hoffmann and I had two days before. The sky was cloudy and the air cold, 25 degrees. We were on top about two hours. We planted the American flag on top, and left a paper in a bottle with our names, the height, etc. It is not at all probable that any man was ever on the top before, or that any one will be again—for a long time at least. There is nothing but love of adventure to prompt it after we have the geography of the region described.

We were back before sundown; a hearty dinner and pleasant fire closed the day. We sang "Old John Brown" around the camp fire . . . we three, alone in these solitudes. Thus was spent Independence Day. (The last was with Hoffmann alone, in the Sierra farther north.) We heard not a gun. Would that we might know the war news—we are over a month behind. [L]

Gardiner's observations from the summit are quoted verbatim in the Report: "Canyons from two to five thousand feet deep, between them ridges tipped with pinnacles sharp as needles; successions of great, crater-like amphitheaters, with crowning precipices over sweeping snow-fields and frozen lakes; everywhere naked shattered granite without a sign of vegetation, except where a few gnarled and storm beaten pines cling to the rocks in the deeper canyons; such were the elements of the scene we looked down upon, while cold grey clouds were drifting overhead."

The note left in a bottle by Brewer and Gardiner on the summit was found in 1896 by Joseph N. LeConte and was eventually taken to the Sierra Club in San Francisco for preservation. It was destroyed in the 1906 earthquake and fire. (A photograph of the note is reproduced in Farquhar's Up And Down California.*)*

TUESDAY, JULY 5

Night very clear and sky black. We were so tired that we laid in our blankets 11 hours. A most lovely morning. [Took] latitude observations and wrote. A cloudless day. Air mild and balmy, sky deepest blue black and rest after our yesterday's climb is very gratif[ying]. [D]

WEDNESDAY, JULY 6

Packed up and came down into the canyon of the South Fork of the Kings to a fine grassy meadow about 2000 feet below our last camp, the last 1000 feet down the steep side of a gigantic moraine, the hardest trail yet reached. At sunset a most lovely rosy tint tinged the granite peaks above. [D]

Brewer, Hoffmann, and Gardiner left Camp 171 in search of better forage. Leaving a note for King and Cotter, they de-

scended to a meadow closer to Roaring River. The new camp was Camp 175. Camps 172–174 were those established by King and Cotter.

Thursday, July 7

Slept until very late. In camp all day, writing, reading and observing, the dinner taking some time as short rations make more care in making a little go a long way. A clear lovely day and again the most glorious tints on the peaks, between a rose and salmon color. [D] For three days the sky has been of the intensest blue, not a cloud in sight day or night. [L]

Friday, July 8

Just before night the joyful shouts of Dick and King—they were successful. Climbed the peak they started for, found that it was not the highest. Had a terribly rough time. Camped at 12,000 feet. Climbed heavy precipices. Had a very rough time of it—we are relieved that they are safely back. [D]

They got on a peak [Mt. Tyndall] nearly as high as Mount Shasta, or some 14,360 feet, and saw five more peaks still higher. They slept among the rocks and snow one night at an altitude of twelve thousand feet, crossed canyons, and climbed tremendous precipices, where they had to let each other down with a rope that they carried along. It was by far the greatest feat of strength and endurance that has yet been performed on the Survey. The climbing of Mount Shasta was not equal to it. Dick got his boots torn off and came back with an old flour sack tied around his feet. [L]

Based on King's notes, his account in the 1865 report, and his longer narrative in Mountaineering, *the historic path of King and Cotter can be reasonably reconstructed. The four ascended Brewer Creek Canyon to the base of Mt. Brewer, turned right and entered a broad u-shaped pass on the southwest spur of Mt. Brewer, faced left and then climbed east to the saddle between Mt. Brewer and South Guard Peak. This elevation (12,800 ft.) disclosed a vista of successive blades and gorges ahead of the pair. King and Cotter decided to try to stay high and climb south along the ridge line of the "Mt. Brewer wall," the Great Western Divide, all the way to the Kings-Kern Divide, the sharp ridge crossing to the chain of even taller peaks to the east. They ascended the next peak to the south of Mt. Brewer, South Guard. They saw that the way ahead was blocked by "sharp needles." Consequently, they descended the east side at Longley Pass (12,500 ft.), now a knapsack route unjustly rated as "very easy" by the* Climber's Guide. *This deposited them at a frozen lake (Lake 11,459) in an "amphitheater." They dropped yet farther (toward Lake Reflection) until they found a way back up to "another amphitheater" adjacent to the Kings-Kern Divide. They proceeded south to "the last alcove of the amphitheater, just at the foot of the Mt. Brewer wall" near two frozen lakes, camping on a granite shelf near one of the two lakes.*

After rising at 4 a.m. and breakfasting on venison by starlight, they climbed southward toward a niche in the Kings-Kern Divide, "which bounded us only half a mile distant," cutting steps in the steep frozen snow with Cotter's bowie knife. They crossed near Mt. Jordan. Seeing that the way below was comparatively flat, they abandoned the divide and descended to a lake "one half mile long." On flat terrain, they strolled to a "little grove of pines" at approximately 11,000 feet, very likely along Tyndall Creek, where they camped.

The next day, the pair ascended what they believed was the highest point in the range and had already named Mt. Whitney in honor of their chief. At the top, however, they saw several higher peaks. Significantly, they named the highest "Mt. Grant." King's field book notes compass bearings "from Mt. Whitney," now Mt. Tyndall, of other peaks observed during this ascent. "Mt. Grant" is shown at "S 42 E," (magnetic bearing) and "300 feet higher." This closely fits Mt. Whitney. (Mt. Tyndall is 14,018 ft. and Mt. Whitney is 14,494 ft.) The Grant in question was, no doubt, Ulysses Grant, who had collected a series of Union victories in 1863–64, had been placed in charge of all Union armies earlier in 1864, and, at the time King and Cotter stood atop Mt. Tyndall, was hailed, according to Carl Sandburg, as the most popular man in the United States.

When King and Cotter rejoined the group, however, Whitney's name was attached to the highest peak; Tyndall was given the honor of King and Cotter's conquest; and Mt. Grant sank without trace in Sierra history. So it was that there was a brief period when the highest point in the United States was Mt. Grant, not Mt. Whitney. Two other place names that have survived were Milestone and Table Mountains on the Great Western Divide, also noted in King's field book.

The return trip was by a somewhat different route (the daring descent of the Kings-Kern Divide being impossible to reverse) and is more difficult to reconstruct. It is possible that King and Cotter re-crossed the Kings-Kern Divide via Milly's Foot Pass and camped near Lake Reflection and then returned via Longley Pass. King specifically recalled that they finally reached the same shoulder on Mt. Brewer whence they had started. When they descended to Camp 171, a note directed them to the new camp below.

"That night," Brewer wrote in his January 11, 1873, letter honoring Professor Tyndall, "by the light of the campfire, I calculated the height so well as possible with only their observations. It was the highest unnamed peak then measured in the country. We called it Mt. Tyndall."

SATURDAY, JULY 9

Up in good season and started back [to Camp 168 near JO Pass]. Tried a cutoff at the last part which proved as usual much the longer. Saw several deer and just at evening Dick shot one, a fine one, much to our relief, for we were reduced to jerked bear and beans. We got a late supper of fresh deer liver and resolved to stay over Sunday here. [D]

Brewer made his final entry in his letter to Professor Brush on the 9th, summarizing: "We are on our return from finding the grandest group of mountains perhaps in the United States. Five peaks over 14,000 and possibly as high as Mt. Shasta with perhaps 50 points above 13,000 ft. King performed a great feat

12 LONGLEY PASS

King and Cotter had planned to stay high on the Great Western Divide until it reached the Kings-Kern Divide. The route, however, proved impassable at some pinnacles and they were forced to descend the Great Western Divide to the east at a point north of the pinnacles.

In our 1986 exploration, we passed through the notch shown in Plate 11. The crest of the divide is a few hundred yards east of the notch. On July 4, 1864, the four parted company at this point on the divide, with King and Cotter proceeding up or around South Guard. We contoured around South Guard. On its south side is a broad saddle. Further south the divide becomes impassable. The saddle spills over the east side into a steep canyon that includes Lake Reflection and Lake 11459 at a cross-country gap now called Longley Pass.

This region is all well over 12,000 feet. It is a land of rock, ice and water. In 1986, we found the east side of the entire pass blocked by a fifty-foot cornice of snow—even in late August. Judging from our observation of the terrain and King's notes, this pass was almost certainly the same one used by King and Cotter.

of mountaineering by getting in five days out farther than any of the rest of us could go."

SUNDAY, JULY 10

We are on our way back for more provisions and on short rations. [L]

A lovely day. About noon, two soldiers came to camp looking for us. They had been at Big Meadows for several days, and we not returning, they struck our trail and came on. Heavy toothache set in and I spent a miserable night. [D]

MONDAY, JULY 11

Went to Big Meadows with a roaring toothache all the way. Found the soldiers there with a pile of newspapers, the first we have seen for a long time. [D] An escort of seven soldiers had been there several days waiting for us. They were having a good time and were eating venison at a heavy rate. [L]

13 ROCKS AND FLOWERS, LAKE 11459

King wrote that their descent from the divide deposited them at a frozen lake in an amphitheater. This was either Lake 11459 or the smaller one a few hundred feet above it, which still had ice patches when we found it in late August 1986. From Lake 11459 they continued to descend toward Lake Reflection until they found a route up into yet another amphitheater in the corner of the Great Western Divide and the Kings-Kern Divide.

Once our group made it to Lake 11459, we stopped for the day, engulfed in mist and rain. Occasionally, the gorge below or the pass above broke through the fog. Along the way to the lake, the first green we found was a fine field of Shooting Stars.

TUESDAY, JULY 12

I started for Visalia, sixty miles distant, to have [the tooth] out. King went down in order to take another trail and reach the high peaks and region which had been inaccessible to us. [L] Gardiner and Dick start[ed] for supplies which are at the mill. [D] At twenty-five miles we struck a pleasant ranch, in a little valley, where we stopped all afternoon. It was a nice place, and we got two very nice meals—the first square meals for some time and we did them ample justice. It was hot, we dared not ride farther by day, but just at sundown we were off again. It was moonlight until midnight, but we rode all night and got into Visalia just after sunrise. [L]

WEDNESDAY, JULY 13

Just before daylight the ulceration in my jaw broke, and what a relief it was. My face was badly swollen, and for over fifty hours it had been terrible—by far the worst toothache I had ever had. We got our breakfast, then went to bed, and slept until noon. It was intensely hot, and we felt it, coming from the cool mountains. [L] I got the tooth out and attended to some business. Peaches and melons are abundant and we made fine use of them. [D]

From Visalia Brewer sent an account of the explorations to date to Whitney, who replied on August 10 (from Northampton, Massachusetts): "In the first place let me congratulate you and the party on the highly important results you have achieved. I had been very much afraid that this expedition would turn out to be a comparative failure, owing to the drought and I am greatly pleased that you have made such a brilliant beginning. As for King, he deserves a gold medal for his pluck. The discovery of a new pine [Pinus Breweri] *will be a very elongated feather in your cap."*

THURSDAY, JULY 14

I got an escort of two men from the camp to assist King [L] but could not get a pack-mule. [D]

I started back alone. I rode about twenty-five miles before the moon went down. I had got into the foothills and could not see my way, so I pulled off the saddle and lay down under a tree and slept two hours. [L]

On the 14th, King started for Mt. Whitney via a more southerly route. He later described the plan to Whitney as follows: "My instructions from Prof. Brewer were to make a strong effort to reach Mt. Whitney, to secure all the topography I could in such a rapid trip (and with a pocket compass whose needle was bent) and to report to Clark's Ranch on August 1st. To accomplish this he gave me one hundred dollars, out of which was to be paid the bill for ourselves and animals at Visalia. I suggested two plans for rejoining the party. One was to meet them at any given point either on the summit, or in Owens Valley, where I thought they must go; secondly, to rendezvous at Millerton, where I knew they must send for supplies; both of these proposals (as facts afterwards proved) were not as absurd as they were pronounced, for on either plan I should have rejoined the party a month before I did and when the working force of the party was at low ebb." (King's field notes indicate

14 EAST LAKE AND THE KINGS-KERN DIVIDE

East Lake lies immediately east of Mt. Brewer and a few miles north of the Kings-Kern Divide. King and Cotter scaled the Kings-Kern Divide in order to reach Mt. Tyndall, although they initially did so at a point west (right) of the scene shown. It is possible that on their return trip they crossed the divide via Lucy's Foot Pass (in the center of the photograph), although their return route is unclear. East Lake was noted by Brewer in his observations from the summit of Mt. Brewer.

15 MT. LANGLEY FROM MT. WHITNEY

King and Cotter walked into Sierra history in their famous try for Mt. Whitney, but they had to settle for Mt. Tyndall and the Kings-Kern Divide. Then King tried it from a more southerly route. On July 23 at noon, King resigned within 400 to 800 feet of the summit. In 1871, King climbed Mt. Langley, thinking it was Mt. Whitney. In 1873, he finally scaled Mt. Whitney.

The ascent of Mt. Whitney is today much easier, as a trail extends to the top. In 1981, Kirby Wilcox, Bill Helvestine, good hiking friends, and I entered at Kearsarge Pass and followed the Muir Trail over the Kings-Kern Divide at Foerster Pass, the highest north-south pass in the Sierra, and then struck west to Milestone Basin. We climbed Milestone, found our way back to the Muir Trail and gradually followed the trail to the western side of Mt. Whitney. Our final morning began at Guitar Lake. At dawn, we slowly headed up the mountain, singing songs, to Trail Crest (13,700 elevation). We temporarily left our packs there and continued on with day packs to the summit of Mt. Whitney. This scene looks from the summit to Mt. Langley.

that, in returning, he nonetheless proceeded to Clark's via Millerton, no doubt in hope of finding and returning with the squad of soldiers Brewer ultimately did send to Millerton for provisions.)

The 1865 Report states that King left Visalia on the 14th with an escort of two soldiers and proceeded east on the Owens Lake and Visalia Trail, a road then being constructed by a local cattleman named John B. Hockett. The trail was not completed beyond the Kern River. From there, for several days, King hiked east and found Mt. Whitney. By July 23 at noon, according to a letter from Brewer to Whitney on March 23, 1865, King had climbed to within 400 to 800 feet of the summit but was unable to reach the top. King wrote Whitney: "When I hung the barometer, a bitter wind was blowing, it hailed and snowed and the clouds closed in around me shutting off all view. I had been all day alone and the fatigue and excitement exhausted me greatly, so much that although chilled through I could scarcely prevent myself from lying down and sleeping. A strange carelessness came over me, making me reckless in the descent. I was obliged to make an effort to keep myself from running over the most dangerous debris slopes."

King plainly located the correct mountain, for he reported that "southward of the main peak, there is a range of sharp needles, four of which are over 14,000 feet high." This referred to the Keeler Needles, the southern companions to Mt. Whitney. King also reported, correctly, that Mt. Whitney has "the outline of a helmet, the perpendicular face being turned toward the east." King returned to Visalia and later rejoined the rest of the Survey upon their arrival at Clark's Station in August. Based on King's measurements, the 1865 Report proclaimed Mt. Whitney "the highest point in the United States."

Friday, July 15

When day dawned I went on. I got my breakfast at a miserable cabin where I had a vivid idea of what stuff some people can live on. I got to Lewis' Ranch, the nice place I spoke of, about nine or ten o'clock, and staged there all day and night, as it was too far to go on that day. [L]

Saturday, July 16

Came on early and struck up the trail. I missed it once and for two hours worked my way through the dense brush to a hill where I could see where the trail was. My horse got away, getting frightened. Luckily, he came back to me. Found the trail and was safely in camp long before night. [D] Hoffmann and Gardiner ha[ve] been to Thomas' Mill and ha[ve] obtained the provisions we had there. [L]

Kings Canyon

Sunday, July 17

Packed up and came on about seven miles by a horrible trail to some meadows near the middle fork of the Kings River. Once a mule fell down some rocks. I thought him killed but he got up again with some bad cuts and bruises. Found an old hunter's or Indian camp where we stopped and the boys saw a bear but could not get him. One of the boys while hunting came suddenly on a man [from the Stebbins party, a group exploring the same region], and found that they were camped four miles down the trail. [D]

There are seven soldiers with us, fine fellows, who are right glad to get out of the hot camp at Visalia. They are mounted, armed with Sharp's carbines and revolvers, and have a month's rations on three pack mules. We made quite a cavalcade—eleven men and sixteen animals—and left quite a trail. [L]

Monday, July 18

Up and off very early by a terrible way following in the main an old Indian Trail. [D] In about four miles we came on a camp of half a dozen men, prospectors who had crossed the mountains from Owens Valley and worked their way thus far. [L] [This was the] Stebbins party and three of their men came down with us. While on the steep side of a hill, Nell's pack slipped back and she fell, and

16 KINGS CANYON

While King searched for Mt. Whitney, the rest of the Survey descended into Kings Canyon to the north. "Next to Yosemite," Brewer wrote, "this is the grandest canyon I have ever seen. We camped at the head of this valley by a fine grassy meadow where the stream forks. On both sides grand walls of granite about three thousand feet high, while between the fork a stupendous rock, bare and rugged [rises] over four thousand feet high." The latter may have been the Grand Sentinel, the highest point in this scene from the banks of the Kings River, looking southwest.

Kings Canyon is today accessible by automobile (Highway 180 ends in the canyon). It has many parallels to Yosemite Valley and is the choice of many as the second most beautiful valley in the Sierra. It is vastly different in one respect, however. Whereas Yosemite Valley has been developed with hotels, lodges, shops, and the like, Kings Canyon has virtually no amenities. There is only a tiny cafe, a small lodge, and one general store, all in one building.

17 PARADISE VALLEY AND KINGS CANYON

From Kings Canyon, the Survey searched for a northerly exit, hoping they could connect their survey with Mt. Goddard farther north. On July 20, Brewer "explored a side canyon, to the north, where the Indian foot trail ran." This was Paradise Valley, which rises gently from the head of Kings Canyon. This scene is from Paradise Valley looking south to the Sphinx. The head of Kings Canyon, where Camp 179 was located, lies just this side of the Sphinx. From there, the main canyon continues west (to the right).

rolled over and over down the steep bank for 150 feet and then brought up against a log. Part of the pack went and a flour sack bursted and we lost some. Wonderful enough she was not seriously hurt. We got her back to the trail. Carried up the pack and soon she was ready for another start. We sank into the canyon of the South Fork of the King's River down a steep hill for 3000 feet. The view almost rivalled Yosemite. Fine camp [178] under cliffs. Boys caught 38 trout. [D]

Meeting the "Stebbins party," a group of prospectors from the Owens Valley, proved fortunate in two ways. As Brewer's diary indicates, three of the prospectors decided to return to the Owens Valley and showed the way into Kings Canyon via roughly the Don Cecil Trail. In addition, they also explained how to pass out of the canyon along Bubbs Creek and over Kearsarge Pass to the Owens Valley, as the Survey eventually was forced to do.

TUESDAY, JULY 19

A sumptuous breakfast on trout. Ate eight apiece, then up the canyon 10 miles, the grandest ride of the year. The Valley resembles Yosemite in places, the cliffs rising 3000 feet and for the most part way over 2000 feet, often nearly vertical. [D]

Next to Yosemite this is the grandest canyon I have ever seen. We camped at the head of this valley by a fine grassy meadow where the stream forks. On both sides grand walls of granite about three thousand feet high, while between the fork a stupendous rock, bare and rugged over four thousand feet high. [L]

Camp 179 was at the upper end of Kings Canyon. The most likely location is just beneath Glacier Monument at the head of the canyon between the forks of the river, one extending into Paradise Valley and the other rising east as Bubbs Creek. The key phrase, however, may be "bare and rugged over four thousand feet high," for only the Grand Sentinel fits that description. This would place the camp near Zumwalt Meadow, not quite at the head of the canyon. The "fork" in the stream may have simply been the fork consisting of the Roaring River and the South Fork.

WEDNESDAY, JULY 20

We started in different directions. Hoffmann and Gardiner climbed the cliffs on the south side. They got up 2000 feet by hard climbing, only to find walls a thousand feet above them which they could not scale. I explored a side canyon [Paradise Valley], to the north, where the Indian foot trail ran, to see if we could get out that way with our animals. I had a grand climb, but found the way entirely inaccessible for horses. I followed up a canyon, the sides grand precipices, with here and there a fine waterfall or series of cascades, making a line of foam down the cliffs. I climbed over boulders and through brush, got up above two very fine waterfalls, one of which is the finest that I have seen in this state outside of Yosemite [Mist Falls]. I had a hard day's work. [L]

The canyon here is very much like the Yosemite. It is a valley, from half a mile to a mile wide at the bottom, about eleven miles long and closed at the lower end by a deep and inaccessible ravine like that below the Yosemite, but deeper and more precipitous. It expands above and branches at its head, and is everywhere surrounded and walled in by grand precipices, broken here and there by side canyons, resembling the Yosemite in its main features. The walls of the Kings River canyon, however, are nowhere vertical to so great a height as Tutucanula; but rather resemble the Sentinel and Cathedral Rocks, or the Three Brothers, of the Yosemite Valley. They rise at various points to heights estimated to be from 3500 to 6000 feet above their base. At the head of the valley, occupying a position analogous to that of the Half Dome at the Yosemite, is the most elevated part of the wall; it is nearly vertical, and between 6500 and 7000 feet high. But it has no features so striking as the Half Dome, or Tutucanula, nor has it the stupendous waterfalls which make that valley quite unrivalled in beauty. [R]

The goal of these explorations was to find a route north to Mt. Goddard, a dark eminence approximately 20 air miles to the north of Kings Canyon. This elevation, which the Survey named in honor of an early California topographer, had been prominent along the northern horizon most of the summer. The Survey desired to reach it in order to obtain the topography of the wide region about Mt. Goddard.

THURSDAY, JULY 21

Left our picturesque camp and came out of the canyon by the way the boys had found [on the north wall], a terrible trail, up just as steep as material will lie for a thousand feet. Camped at last about five miles from our last camp and 4000 feet above it in a most picturesque spot with the grand peaks of the crest [the Brewer Group] in full view. [D]

It was heavy for our animals. Twice we had very steep slopes for a thousand feet together, where it seemed at first that no animal could get up with a pack. Once our pack horse fell, turned a complete somersault over a boulder, and landed below squarely on his feet, when he kept on his way as if nothing had happened. His pack remained firm and he was not hurt in the least. Fortunately it was not so steep there. There were places where if an animal had once started he would have rolled several hundred feet, but all went safely over. [L]

Brewer noted in a letter that the route had been found by one of the soldiers. According to Brewer's field book, the route was westerly down the main canyon and then north up a steep creek. The Hoffmann Map shows the way ascending along Copper Creek in roughly the way taken by today's trail. Brewer's field notes refer to green copper stains on the rock. Camp 180 was probably in or slightly above Upper Tent Meadow along Copper Creek.

FRIDAY, JULY 22

Gardiner and Hoffmann went on a peak about twelve thousand feet [Goat Mt.], which commands a comprehensive view of all the ground we have been over lately; while two soldiers, Dick and I explored ahead for a trail.

We were unsuccessful, but we got on a ridge over eleven thousand feet high that commands a stupendous view. The deep canyons on all sides, the barren granite slopes, clear little lakes that occupy the beds of ancient glaciers, the sharp ridges, the high peaks, some of them rising to above fourteen thousand feet, like huge granite spires—all lay around, forming a scene of indescribable sublimity. [L] [Hoffman and Gardiner] think they saw a trail and way north. [D]

The crest presents a very serrated outline. Two peaks lying just in front of it are especially fine; they are between five and six miles east of Camp 180; both are probably over 14,000 feet high, the northern being a little the highest. This we named Mount King, and the southern one Mount Gardiner. [R]

The southern tip of Mt. King was later named Mt. Cotter.

SATURDAY, JULY 23

We thought the region north impassable for our animals, but Hoffmann and a soldier both saw another way they thought practicable, and three have gone today to explore it. If that too should prove impracticable we shall be in a hard fix and will have to make our way to Owens Valley and cross the Sierra again at some point north. [L]

Wrote most of the day. Boys returned unsuccessful, so we are resolved to take back trail. [D]

The Survey was unsuccessful in its attempts to find a passable route to the north from the Monarch Divide. The Hoffmann Map indicates they found but did not go beyond Granite Pass. Consequently, as Brewer wrote his brother on August 5, "I decided to cross the summit to Owens Valley." This required that the group return to Kings Canyon and then push east, up Bubbs Creek, and then over the crest at Kearsarge Pass—along the route of the prospectors.

SUNDAY, JULY 24

We remained in camp for latitude observations, but the forenoon was cloudy and the afternoon rainy—heavy showers. [L] Cleared up at evening and turned in my wet blankets, for I had got them wet in keeping the instruments dry. [D]

MONDAY, JULY 25

I had rheumatic night in my wet blankets. There were light showers during the night, but no heavy rain, and the morning was clear. We packed up, got back into the canyon by our steep trail, killed a tremendous rattlesnake on the way, and camped again at the head of the valley, where we had a week before [Camp 179]. One of the soldiers caught a fine mess of trout. We have seen deer in abundance, but have not succeeded in getting any lately. [L]

TUESDAY, JULY 26

We started and got about eleven miles [ascending Bubbs Creek], a hard day's work, for we rose 4,300 feet. First we went up a steep, rocky slope of 1,000 to 1,500 feet, so steep and rough that we would never have attempted it had not the prospectors already been over it and made a trail in the worst places—it was terrible. In places the mules could scarcely get a foothold where a canyon yawned hundreds of feet below; in places it was so steep that we had to pull the pack animals up by main strength. They show an amount of sagacity in such places almost incredible. Once Nell fell on a smooth rock, but Dick caught her rope and held her—she might have gone into the canyon below and, with her pack, been irretrievably lost. We then followed up the canyon three or four miles. We camped by a little meadow, at over nine thousand feet. Near camp a grand smooth granite rock rose about three thousand feet, smooth and bare [Charlotte Dome]. [L]

WEDNESDAY, JULY 27

We went over the summit [at Kearsarge Pass], about twelve miles [from Kings Canyon]. The summit is a very sharp granite ridge, with loose boulders on both sides as steep as they will lie. It is slow, hard work getting animals over such a sliding mass. It is 11,600 feet height, far above trees, barren granite mountains all around, with patches of snow, some of which were some distance below us—the whole scene was one of sublime desolation. Before us, and far beneath us lay Owens Valley, the desert Inyo Mountains beyond, dry and forbidding. Around us on both sides were mountains fourteen thousand feet high, beneath us deep canyons. [L]

Then down as steep a trail and down the canyon of Little Pine by a horrible trail. Bold were the men who first went through. We appear to be the second. Passed a number, some six to seven, lovely lakes. [D]

We camped at a little meadow [Onion Valley] in full view of the valley below and the ridges beyond, which were peculiarly illumined by the setting sun. [L]

The most practical route, based on today's knowledge of the terrain, would have been to follow Bubbs Creek to Vidette Meadow and thence to the pass. The Hoffmann Map, however, shows the route leaving the main canyon near Charlotte Dome, well before Vidette Meadow, and following Charlotte Creek to Charlotte Lake and thence to the pass. This is consistent with Brewer's statement to his brother that they "followed up the [Bubbs Creek] canyon three or four miles and then out by a side canyon still steeper," with his diary entry for August 26 that the side canyon was "to the north" and "terribly steep," and with the fact that the Hoffmann Map is more accurate in its detail of the Charlotte Lake corridor than the Vidette Meadow route.

The Owens Valley

THURSDAY, JULY 28

We were up at dawn and went to Owens Valley, 16 miles. Six miles brought us out of the canyon on the desert—then ten miles across the plain in the intense heat, and we camped on the river bank, without shade or shelter, the thermometer 96 degrees in the shade, 156 degrees

18 MIST FALLS

Of his exploration of Paradise Valley Brewer wrote: "I followed up a canyon, the sides grand precipices, with here and there a fine waterfall or series of cascades, making a line of foam down the cliffs. I climbed over boulders and through brush, got up above two very fine waterfalls, one of which is the finest I have seen in this state outside of Yosemite." This was Mist Falls in Paradise Valley, here shown at high water in late Spring 1986.

19 KINGS CANYON GORGE AND THE BREWER GROUP

Searching for a northerly exit from the canyon, the Survey climbed along the Copper Creek drainage to the crest of the Monarch Divide. This is the view from near the divide looking south into the canyon and across to the Brewer group and the Great Western Divide. Camp 180 was established near the top of the drainage in view of Kings Canyon on July 21.

Today the same region can be reached via the Copper Creek Trail from the floor of Kings Canyon. This is one of the steepest trails in the Sierra with many switchbacks. In late September 1985, I started up the trail. The aspen had turned and water was low. As the afternoon wore on, no flat and decent camp presented itself, so I went on over the divide into Granite Basin, more than a vertical mile from trailhead! Hoffmann's map shows their route was very much the same, reaching to Granite Pass. The next day I crossed back in view of Kings Canyon and tried to determine where Camp 180 might have been. My best judgment is that the camp was in or near the meadow in the foreground of this photograph. As Brewer noted, there is a fine vista of the Brewer Group from this camp.

in the sun. Yesterday in the snow and ice—today in this heat! It nearly used us up.

Owens Valley is over a hundred miles long and from ten to fifteen wide. It lies four thousand feet above the sea and is entirely closed in by mountains. On the west the Sierra Nevada rises to over 14,000 feet; on the east the Inyo Mountains to 12,000 or 13,000 feet. The Owens River is fed by streams from the Sierra Nevada, runs through a crooked channel through this valley, and empties into Owens Lake twenty-five miles below our camp. This lake is of the color of coffee, has no outlet, and is a nearly saturated solution of salt and alkali.

The Sierra Nevada catches all the rains and clouds from the west—to the east are deserts—so, of course, this valley sees but little rain, less than three fourths of an inch in 1861–62, but where streams come down from the Sierra they spread out and great meadows of green grass occur. Tens of thousands of the starving cattle of the state have been driven in here this year, and there is feed for twice as many more. Yet these meadows comprise not over one-tenth of the valley—the rest is desert. At the base of the mountains, on either side, the land slopes gradually up as if to meet them. This slope is desert, sand, covered with boulders, and supporting a growth of desert shrubs.

Here is a fact that you cannot realize. The Californian deserts are clothed in vegetation—peculiar shrubs, which grow one to five feet high, belonging to several genera, but known under the common names of "sagebrush" and "greasewood." They have but little foliage, and that of a yellowish gray; the wood is brittle, thorny, and so destitute of sap that it burns as readily as other wood does when dry. Every few years there is a wet winter, when the land of even these deserts gets soaked. Then these bushes grow. When it dries they cease to put forth much fresh foliage or add much new wood, but they do not die—their vitality seems suspended. A drought of several years may elapse, and when, at last, the rains come, they revive into life again!

The Inyo Mountains skirt this valley on the east. They look utterly desolate. They were the stronghold of the Indians during the hostilities of a year ago. They are destitute of feed, and the water is so scarce and obscure that the soldiers could not follow them without great suffering for want of water. For a year fighting went on when at last they were conquered. One chief, however, Joaquin Jim, never gave up. He retreated into the Sierra with a small band but he has attempted no hostilities since last fall.

We camped on the river near Bend City [a boom town then on the east bank of the Owens River but long since vanished] and went into town for fresh meat and to get horses shod. It is a miserable hole, of perhaps twenty or twenty-five adobe houses, built on the sand in the midst of the sagebrush, but there is a large city laid out—on paper. It was intensely hot, there appeared to be nothing done, times dull, and everybody talking about the probable uprising of the Indians—some thought that mischief was brewing, others not. [L]

In 1862 there had been numerous killings between settlers and Paiutes in Owens Valley, leading to the establishment by the army of Camp Independence on July 4th of that year. Within a year, a peace treaty was made and the camp abandoned. Many Paiutes moved to a reservation close to Fort Tejon, near Bakersfield, but Joaquin Jim and a band of followers stayed in the northern valley and the mountains. The worst of the fighting was over when the Survey arrived in 1864. The following year, however, hostilities broke out again and Camp Independence was re-established.

FRIDAY, JULY 29

We were up at dawn and started up the valley, and travelled twenty-two miles, about three-fourths of the way over deserts, the rest over grassy meadows. Many settlers have come in with cattle this year. Most of them Secessionists from the southern part of the state. We passed the old Camp Independence, so lively a year ago—now a ranch, and the adobe buildings falling into ruin. [L]

20 MTS. GARDINER AND BREWER FROM MT. COTTER

The names of all five members of the expedition grace Sierra peaks. Four stand close by one another in Kings River country. The southern tip of Mt. King was eventually named for Cotter. This scene is from that tip, looking southwest to Mt. Gardiner and Mt. Brewer (in the distance and left). Mt. Clarence King is behind the camera. (Mt. Hoffmann, named in 1863, is much farther north in Yosemite National Park.)

In 1984, Kirby Wilcox, Bruce Dodge, and I visited this region. We entered via Kearsarge Pass and soon turned north on the Muir Trail. We crossed Glen Pass and dropped into the Rae Lakes Basin, a popular spot. We took a day hike into Sixty Lakes Basin to climb Mt. Cotter. This photograph was taken from just below the summit block.

Camped at Black Rock, the scene of fearful Indian tragedies. All used up with the heat. Took a refreshing bath in the river and in our blankets early regardless of the Indians who are expected to rise. [D]

Saturday, July 30

We kept on our way up the valley. It was not so hot, 100 to 104 degrees, most of the time 101 degrees. We had the same features—deserts most of the way, grassy meadows where streams came down from the Sierra and spread on the plain, the barren Sierra on the west, the more barren Inyo Mountains on the east. The eastern slope of the Sierra is here almost destitute of trees, save in the canyons and along the streams.

We camped by a creek, where also a stream of warm water comes from some copious hot springs about a mile distant [probably Keough Hot Springs]. We took a luxurious bath. [L] Indian sweat houses near. Indians have been watching us. [D]

Sunday, July 31

We were up very early, but not refreshed, for the mosquitoes had allowed us but little sleep. We went up the valley sixteen or seventeen miles where a lava table crosses the entire valley, barren in the extreme. Here a stream comes down, and there is a sort of basin where there are nine or ten square miles of the best grass I have seen in the state [Round Valley]. Three or four settlers have come in this year with cattle and horses, but there is feed for ten times as many. One has started a garden, to sell vegetables in Owensville, ten or fifteen miles distant, and Aurora, sixty-five miles distant [both are now ghost towns]. [L]

The Owens River flows down on the western side of this lava table, and in places has excavated in it a canyon a hundred feet or more deep. Through here the meadows [are] finer than any seen in the State, the grass being as high as the horses' backs, and the whole being easily kept irrigated by the streams from the Sierra. [R]

21 BUBBS CREEK

Unable to find a passable route to the north, the Survey returned to Kings Canyon and turned east toward the Owens Valley along a beautiful water course called Bubbs Creek, named for one of the prospectors who had shown them the way. Bubbs Creek flows from the crest of the Sierra to Kings Canyon, where it joins the South Fork. The Survey followed Bubbs Creek upstream from that juncture to Charlotte Creek. Today, the trail from Kings Canyon to Kearsarge Pass runs along Bubbs Creek (although it does not follow the Charlotte Creek Drainage). In 1981, my wife Suzan and daughter Allison and I rested at this hospitable spot on our way back from East Lake to Kings Canyon.

Mono Pass to Mount Goddard

MONDAY, AUGUST 1

We were up early. Some of the horses strayed and two hours were spent in finding them so we got a late start. We struck into the mountains by a blind Indian trail. Last year some soldiers were piloted over the summit by friendly Indians, in pursuit of hostile ones, by this trail. One of [the soldiers] was with us and knew the way. Our way was continually up and we had grand views of the desert plain below and the desert mountains beyond. The Sierra lay ahead, with high peaks and cool snows.

We camped in the canyon of the southwest fork of Owens River [Rock Creek] at over eight thousand feet [Camp 187]. As soon as we halted signal smokes rose from hills around, and after dark signal fires blazed, the Indians telegraphing our progress. The fires were bright, but short. [L]

TUESDAY, AUGUST 2

Ice formed last night. Temperature 29. We are up 8500–8800 feet and the change is great. [D]

We were off in good season. As soon as we started the signal smokes again showed that we were watched. I will not repeat, but suffice to say that all of our movements up to the present time have been thus telegraphed. As soon as we stop, smokes rise, when we start they appear, and at night their blaze is seen on the heights—so the Indians know all of our movements.

We crossed the summit. As we approached it, it seemed impossible—great banks of snow, above which rose great walls and precipices of granite—but a little side canyon, invisible from the front, let us through. The pass is very high, nearly or quite twelve thousand feet on the summit. The horses cross over the snow. So far as I know it is the highest pass crossed by horses in North America. [L]

The views from the high points above the trail at the

22 CHARLOTTE DOME

On the evening of July 26, the Survey stopped along Bubbs Creek near a sheer rock wall. "We camped," Brewer wrote, "by a little meadow, at over nine thousand feet. Near camp a grand smooth granite rock rose about three thousand feet, smooth and bare." This was Charlotte Dome, here seen from the east along the Muir Trail to Glen Pass.

23 THE OWENS VALLEY

On July 27, the party crossed over the crest at Kearsarge Pass, and the Survey saw the vast valley between the Sierra and the White Mountains. This is the view (from south of Kearsarge Pass looking to the southeast) of the Alabama Hills, the valley, and Owens Lake, now a dry bed due to diversion of the Owens River to Los Angeles.

summit were of the grandest description. Eight miles to the north was a group of dark, crimson-colored peaks [the Red Slate Range], and twenty-five miles farther in that direction were the snow-clad ranges near Mono Lake [Mts. Banner and Ritter and the Minarets]. In a southerly direction rose a vast mass of granite peaks and ridges [the Abbott group], with the same sharp serrated crests, vertical cliffs overhanging snow-fields and amphitheaters with frozen lakes, which were the main features of the views in the region about the head of [the] Kings River. [R]

We came down a very steep slope 2,500 feet and camped [at an old Indian camping place] on the Middle Fork of the San Joaquin [Mono Creek], near its head. [L] Met eleven Indians, well armed, after camping and one came into camp [188]. [D]

The pass crossed was Mono Pass, not to be confused with the pass by the same name farther north at the top of Bloody Canyon in Yosemite National Park. The Hoffmann Map shows the trail proceeding from Round Valley directly over Wheeler Crest into the canyon of Rock Creek on a line very close to the Mono-Inyo county line.

WEDNESDAY, AUGUST 3

Hoffmann, a soldier, and I climbed a ridge south while Gardiner and another soldier climbed a peak north [the Red Slate Range]. We had a steep climb, and a grand view from the summit, which was over 12,000 feet but it was cut off by the higher crest southeast and south of us [the Mt. Abbott group]. Day perfectly cloudless. The dust and smoke shut out the plain and the Coast Ranges. Back in good season. They did not get back until nearly dark. [D]

[The Red Slate Range] forms the northern termination of the great elevated range of the Sierra, which stretches to the south, for a distance of over ninety miles, without any depression below 12,000 feet, in all probability the highest continuous mass of mountains in North America. To the north, between the Slate Peaks and the

Mono Group, a considerable depression exists, over which is a pass [Mammoth Pass], of the height of which we have no positive knowledge. [R]

THURSDAY, AUGUST 4

We came down the valley [along Mono Creek] to this camp [189], 18 miles. In places the canyon widens into a broad valley. There are many beautiful spots, but they have been rarely seen by white men before. It is the stronghold of Indians; they are seldom molested here, and here they come when hunted out of the valleys. We saw their signs everywhere; their fires and smokes on the cliffs near showed their presence, but we saw not a man, woman, or child. Once we came on a camp fire still burning, but the Indians were out of sight.

In this valley hundreds of pine trees have the earth dug around them to protect them from fire, for pine seeds or nuts form an important article of food with the Indians. One species has very large cones, with large seeds—hundreds of bushels of seeds are gathered for food. [L]

In passing from Camp 188 near the summit down the deep straight cut of Mono Creek, the Survey noted, in the Report, a "prominent peak" of slate to the south, which they named Mt. Gabb in honor of William Gabb, the paleontologist of the Survey (who was not present on the expedition). The name was later moved to the present Mt. Gabb and the identity of the original namesake is not clear. Camp 189 was in Vermilion Valley, so named in 1894, by Theodore Solomons, a beautiful valley since flooded to form Lake Thomas Edison.

FRIDAY, AUGUST 5

This morning four soldiers left for Fort Miller [in the foothills along the San Joaquin north of Fresno] for supplies. They took an unknown way, following the foot trail, but hope to be back in seven days, while we will work south into the region [about Mt. Goddard] we could not penetrate from Kings River. I have stopped here today to wash clothes, mend, write up my notes, etc., for we have been on the go incessantly for the last eleven days. [L]

This marked the beginning of a southward thrust to Mt. Goddard. Brewer explained: "It was my desire to get on this, as it commands a wide view, and from it we could get the topography of a wide region" [L] *and could "connect with the work on the other side of the Kings."* [R]

SATURDAY, AUGUST 6

To 190 down creek two miles, then south, up a very heavy hill. Camped at a pleasant place at above 8000 feet. Saw a dozen deer but none were shot. [D] In going from Camp 189 to 190, the Middle and South Forks of the San Joaquin [Mono Creek and the South Fork] were crossed, and a due south course was kept towards a high point on the ridge, eight miles distant. [R]

SUNDAY, AUGUST 7

A quiet Sunday in camp. [D]

MONDAY, AUGUST 8

Up at earliest dawn and off before five for a peak, south, a little over 10,000 feet. Fine view but smoke came on and shut out the distant landscape. Took observations until nearly noon. View lovely. Cool air, balmy. The plain hid in its veil of haze, the top of which is level and well defined. [D]

Brewer's handwritten list of camps places Camp 190 "near" the South Fork of the San Joaquin but gives no further data. The Report states it was about 30 miles from Mt. Goddard and that the point ascended was 10,711 feet high. The Hoffmann Map shows the trail crossed the San Joaquin midway between Mono Creek and Bear Creek and ascended in a long traverse up Kaiser Ridge. The point climbed was probably Mt. Givens (10,648 ft.) or Ian Campbell (10,616 ft.). Brewer's scribbled notes in his field book state that the top was reached at seven and that the vista extended from "the Mono Group" all the way down the crest "to the Whitney Group."

TUEDSAY, AUGUST 9

Went on to 191, a long ride over very rough country.

In one place we worked along a ridge but two rods wide at the top and part of that rising in bare rocks while both sides of this ridge were so precipitous that one could see the green meadows 2000–3000 feet beneath. At last struck the branches of the Kings River and camped at the head of one near 10,000 feet. [D]

Camp 191 was intended as a base camp to climb Mt. Goddard. The Hoffmann Map shows the route along Kaiser Ridge, passing the divide into the Kings River drainage near Johnson and Lost lakes (at about where the trail now crosses the divide). Camp 191 was probably at the headwaters of Burnt Corral Creek.

WEDNESDAY, AUGUST 10

Four of us started at 5 o'clock to reach Mt. Goddard, supposed to be seven miles distant. Morning cool and invigorating and we got along well. Ridge after ridge was crossed, each sharp and rocky and most of our way was at 11,000 feet and over. [D] The only possible way led along the divide between the Kings and San Joaquin Rivers. [R]

At 2 p.m., after nine hours heavy walking, we surmounted the seventh ridge to see that the mountain was at least five hours walk farther. Hoffmann had a sore foot and I was nearly given out—so we stopped. Dick and Spratt [one of the soldiers] resolved to make the peak and reached within 300 feet of the summit at 7 p.m. Then back and walked all night. Hoffmann and I divided our lunch into three parts and returned to 11,000 (possibly higher) and set fire to a large dry stump which lay on a naked rock. It was very cold and froze but our fire kept us warm. Our supper was *very* light, a mere morsel. [D]

Excessive fatigue, the hard naked rock to lie on—not a luxurious bed—hunger, no blankets, although it froze all about us—the anxieties for the others who had gone on and were now out, formed the hard side of the picture. [L] This was Camp 192 or "Cold Camp." [R]

But it was a picturesque scene after all. Around us, in the immediate vicinity, were rough boulders and naked rock, with here and there a stunted bushy pine. A few rods below us lay two clear placid lakes, reflecting the stars. The intensely clear sky, dark blue, *very* dark at this height; the light stars that lose part of their twinkle at this height; the deep stillness that reigned; the barren granite cliffs that rose sharp against the night sky, far above us, rugged, ill-defined; the brilliant shooting stars, of which we saw many; the solitude of the scene—all joined to produce a deep impression on the mind, which rose above the discomforts.

Early in the evening, at times, I shouted with all my strength, that Dick and Spratt might hear us and not get lost. The echoes were grand, from the cliffs on either side, softening and coming back fainter as well as softer from the distance and finally dying away after a great length of time comparatively. At length, even here, sleep, "tired nature's sweet restorer," came on. [L]

THURSDAY, AUGUST 11

Notwithstanding the hard conditions, we were more refreshed than you would believe. After months of this rough life, sleeping only on the ground, in the open air, the rocky bed is not so hard in reality as it sounds when told. We actually lay "in bed" until after sunrise, waiting for Dick. They did not come; so, after our meager breakfast we started and reached camp in about nine hours. This was the hardest part. Still tired from yesterday's exertions, weak for want of food, in this light air, it was a hard walk.

At three in the afternoon we reached camp, tired, footsore, weak, hungry. Dick had been back already over an hour, but Spratt had given out. Gardiner and two soldiers, supposing that Hoffmann and I had also given out, had started with some bread to look for us. We shot off guns, and near night they came in, and at the same time Spratt straggled into camp, looking as if he had had a hard time. Dick and he did not reach the top, but got within three hundred feet of it. They traveled all night and had no food—they had eaten their lunch all up at once. Dick is *very* tough. He had walked thirty-two hours and had been twenty-six entirely without food; yet, on the return, he had walked in four hours what had taken Hoffmann and me eight to do. [L]

The route, according to the Report, was along the divide between the Kings and San Joaquin rivers, i.e., *along the western rim of Goddard Canyon, the LeConte Divide. Cotter and Spratt crossed the canyon at its head near Martha Lake. The Report adds that they climbed "within 300 feet of the summit, and hung up the barometer just before it was too dark to see to read it." Computing that altitude as 13,648 feet, the Report placed Mt. Goddard at 14,000 feet, about 400 feet too high. (The Survey's measurements tended to err on the high side by 100 feet or more.) Cotter also reported that the summit of Mt. Goddard was "made up of alternate veins or beds of slate and granite."*

The Report refers to Brewer and Hoffmann's impromptu camp [192] as "Cold Camp." Although Brewer did not specifically place "Cold Camp," it seems clear that it was in Red Mountain Basin near Hell For Sure Lake. The Hoffmann Map shows the trail extending to the saddle on the southeast rim of the basin leading to Goddard Canyon (just east of what is now Arctic Lake). This is also the logical route as seen from the ground and the first point on the hike from which Mt. Goddard could have been seen, thus informing them they had misjudged the distance. (Cotter and Spratt went over the saddle, up the rest of Goddard Canyon, skirted Martha Lake, and ascended the west side of the peak.) The saddle can also be counted as the "seventh ridge," although the number turns on what is counted as a "ridge" and the actual route to that point.

Within the basin, Brewer and Hoffmann's impromptu camp was near "two clear placid lakes" only "a few yards below." The most logical conclusion, based on viewing the scene, is that Brewer and Hoffmann returned to level ground, placed themselves in view of the saddle in the event Cotter and Spratt returned, and camped near two lakes at approximately 11,000 feet elevation. These indicators suggest that Cold Camp was on the granite strip between Hell For Sure and Horseshoe lakes. This strip has a supply, although sparse, of stunted pines and is at 10,880 feet, about the elevation referred to by Brewer. With respect to the

24 HEADWATERS, MONO PASS

After a sweltering trek up the Owens Valley, the Survey re-entered the Sierra at Mono Pass and established Camp 188 on August 2 near the headwaters at the summit of Mono Creek. This scene is of the headwaters near the pass, showing the delicate grass and mountain flowers (here a Shooting Star) that often grace the rims of Sierra watersheds. Our group came over the pass from the Owens Valley in 1980. Our first camp was a few hundred feet below the crest in the same vicinity as Camp 188.

25 THIRD RECESS

The Survey did a lot of climbing from Camp 188. On August 3, Gardiner hiked north into the Red Slate Range and Brewer hiked south along a ridge toward the Abbot Group. This scene shows Third Recess and Fourth Recess, two hanging valleys above Mono Creek (below and out of view) from Pioneer Basin north of Mono Creek. The Abbot Group is in the distance. The ridge climbed by Brewer is probably the center ridge. The closer canyon is Third Recess.

This photograph was made on the evening of the second day of our 1980 outing. We had taken a cross-country route from our camp below Mono Pass to Pioneer Basin. Bruce Dodge assisted me in setting up for this photograph as a storm rolled toward us in the late afternoon. Later in the trip, we climbed Mt. Abbot, shown in the center. Our route was via Second Recess and Gabbot Pass.

echoes noted by Brewer, had Brewer shouted toward the gap at the head of the basin from this spot, nearby Mt. Hutton (along with the opposite wall of the basin) would undoubtedly have caught and reflected their shouts. I regret that during my visit to this spot I did not think to try to duplicate the echoes.

FRIDAY, AUGUST 12

Came back to our pleasant camp [189] in the Valley of the Middle Fork of the San Joaquin [Mono Creek], a long ride. The detail of the escort had got back from Millerton with provisions about an hour ahead of us with papers and provisions. [D] [W]e were out of flour, rice, beans—in fact, had only tea, sugar, and bacon. This was to be our rendezvous with the soldiers, and they had got in only an hour ahead of us, with an abundance of provisions. Not a great variety, but an abundance of salt pork, flour, coffee, and sugar, so we are all right again.

I have stopped here Saturday and Sunday, to rest, wash, and mend clothes, and write up notes. [L]

SATURDAY, AUGUST 13

The soldiers brought back a lot of newspapers from the camp at Fort Miller—papers from the East, from various parts of this state—old many of them, but very acceptable. [A]fter washing my clothes, I spent the rest of the day in reading. There is a sort of fascination in reading about what is going on in the busy world without, in the noisy marts of trade and commerce, in society and politics, in the busy strife of war, of brilliant parties and gay festivities, and sad battles, and tumultuous debate, while we are here in these distant mountain solitudes, alike away from the society and the strife of the world. [L]

SUNDAY, AUGUST 14

A quiet day in camp. Read and wrote most of the day. Thunderstorm about noon and a little rain. Evening clear and lovely moonlight. [D]

The San Joaquin

MONDAY, AUGUST 15

We packed up and started and made a big day's march toward the north fork. We had rather rough going and finished by going down a steep hill about two thousand feet. The north fork runs in a very deep, rocky canyon still a thousand feet deeper. We camped at good grass, but poor water—some pools that remained in a little swamp. Near our camp the canyon is very grand—a notch in the naked granite rocks. During the day we crossed a number of streams bordered by dense thickets of alders through which we had to cut our way with our knives. In one of these, Buckskin, our pack horse, caught a leg between two rocks and bruised it badly. [L]

The objective, as Brewer wrote in the Report, was the "Obelisk Group," i.e., *the Clark Range, in the Yosemite backcountry. Brewer and Hoffmann had observed the group during the summer before and presumably wished to link their 1864 measurements with these prominent peaks. The expedition, however, was unable to reach them. It was not until 1866 that King and Gardiner climbed Mt. Clark, "the Obelisk."*

The main barrier to the Survey's progress northward was the steep canyon of the "North Fork" of the San Joaquin, since renamed the Middle Fork.

TUESDAY, AUGUST 16

We were off as usual, but soon found that we were "in a fix"—great granite precipices descended ahead of us. We turned back, and after much trouble found a very steep, rocky place where we could get down about one thousand feet to the river, after going only about four miles. It was a lovely spot—a little flat of a few acres, with grand old trees, and high, naked granite cliffs around. The river runs through this, entering and leaving it by an impassable canyon. We found the river full of trout, and the boys caught a fine mess. [L]

26 MT. GABB

As the Survey passed westward down Mono Creek they named a peak for paleontologist William Gabb, who had joined them on other field expeditions. The name has since been moved a short distance to this mountain in the Abbot Group.

27 GODDARD CANYON

On August 5, the Survey, at the end of Mono Creek in Vermilion Valley, turned south in an effort to reach Mt. Goddard, a prominent peak that commands a vista of much of the Sierra. Brewer wished to climb it to connect with the work on its south side and to take advantage of its wide view for topographical measurements. On the evening of August 9, the Survey established Camp 191 as the base camp for a long day hike and ascent of the peak. The thrust for Mt. Goddard was along the divide between the Kings and the San Joaquin—the west rim of Goddard Canyon. This scene looks south up Goddard Canyon from a knoll above the confluence of Piute Creek and the San Joaquin. Emerald Peak in the distance hides Mt. Goddard.

One of the soldiers, Win Orr, was ahead [L], not knowing that we had stopped. [D] We tried to stop him but could not. [L]

The [Middle Fork] of the San Joaquin comes down through a very deep canyon, and the wide, open valley-like depression terminates here. This canyon is from 3000 to 4000 feet deep. Two or three miles southeast of this is a most remarkable dome. It rises to a height of 1800 feet above the river, and presents exactly the appearance of the upper part of a sphere; or, of the top of a gigantic balloon struggling to get up through the rock.

The sides of the canyon are very abrupt, and present immense surfaces of naked granite, resembling the valley of the Yosemite. There are everywhere in this valley the traces of former glaciers, on an immense scale, and as the party rose above the canyon on the north, in leaving the river, the moraine on the opposite side was seen very distinctly, and appeared to be at an elevation of not less than 3000 feet above the bottom of the valley. It was evident that the glaciers which came down the various branches of the San Joaquin all united here to form one immense "sea of ice," which filled the whole of the wide depression spoken of above, and left its moraines at this high elevation above the present river-bed. [R]

WEDNESDAY, AUGUST 17

We started again, putting the pack on another animal. We had a tremendous climb to begin with. First a very heavy hill, 3000 feet high, and just as steep as animals can climb. We struck the trail of Orr, the lost soldier, but we soon came to a region where there were cattle, and their abundant tracks prevented us from seeing which way he went, so we camped, and I sent out two soldiers to look for him. They found his tracks in places, but there were Indian tracks also and we feared that they had got him.

On the high ridge, traversed in getting to this camp, were many boulders of lava, which must have been brought from some more northerly point and dropped in their present position by ancient glaciers. The source of these boulders seems to have been near Mount Clark, in the Obelisk Range. The view from the summit of the ridge was a grand one, commanding the whole of the Mount Lyell and Obelisk groups, as well as the main range of the Sierra to the east, where are many dark-colored peaks, apparently volcanic [Mts. Banner, Ritter, Davis and the Minarets]. A very high and massive peak was seen to the east of Mount Lyell, which it nearly equalled in altitude; it was called Mammoth Mountain, and was estimated to be 13,000 feet in height. [R] We found cattle in the woods, and the soldiers shot a beef and we had fresh meat again, a true luxury. [L]

THURSDAY, AUGUST 18

Stayed all day at this camp, looking for Orr but returned without him, although we found his traces. [D]

FRIDAY, AUGUST 19

Went on to the next camp [196] near peak. Not at all well. Weather hazy. [D]

SATURDAY, AUGUST 20

Hoffmann and Gardiner went on peak. I took pills. Had to take an enormous quantity before they would operate so when they did was rather sick. Lay in camp all day. Hoffmann came back down sick with a lame leg that has been troubling him for some days without any known cause. [D]

I resolved to push out. I was getting run down, felt nearly sick, our provisions were low, and Hoffmann was getting rapidly worse. [L]

Although Brewer's diary and letters are vague concerning the location and objective of the excursion along the northern San Joaquin, the Report sheds more light: "Camp 196, a few miles north of 195, was at the base of a prominent peak [probably in the vicinity of Madera Peak], which was supposed to belong to the Obelisk group [Clark Range], for which the party was

28 COLD CAMP, HELL FOR SURE LAKE

Misjudging the distance on their long day hike to Mt. Goddard, Brewer and Hoffmann stopped to make an impromptu camp at 11,000 feet which they called "Cold Camp" on the night of August 10. They slept without blankets beside a burning stump. Brewer termed it a "picturesque scene," writing that "around us, in the immediate vicinity, were rough boulders and naked rock, with here and there a stunted bushy pine. A few rods below us lay two clear placid lakes, reflecting the stars." One of the lakes was very likely Hell For Sure Lake in Red Mountain Basin.

In June 1986, my daughter Allison and I set out to try to find Cold Camp. We drove to Courtright Reservoir on the western slope of the range (northeast of Fresno). From there we took trails to Red Mountain Basin, due east. We made it as far as Hell For Sure Lake. As usual, the terrain suggested the probable route of the Survey of 1864. Cotter and Spratt had little choice but to continue toward the saddle at the head of Red Basin, south of Hell For Sure Lake. This saddle would also have been the first point at which they could have learned that they had misjudged the distance to Mt. Goddard. My best judgment concerning the location of Cold Camp was that it was on the granite strip between Hell For Sure Lake and Horseshoe Lake. This scene shows a snowfield sinking into the icy depths of Hell For Sure Lake.

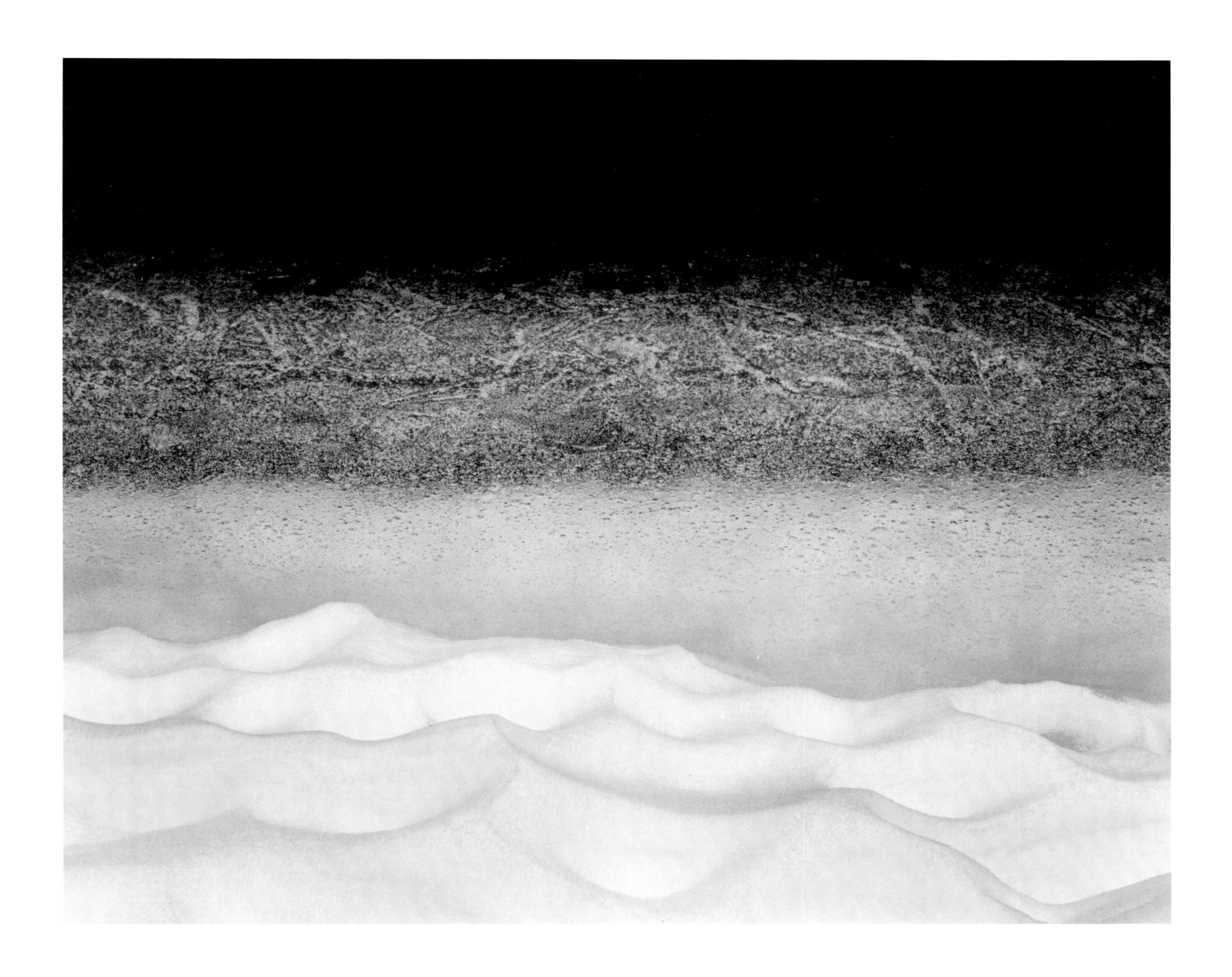

29 MT. GODDARD, WANDA LAKE

Cotter and Spratt, a soldier, went on and came within 300 feet of the summit of Mt. Goddard. On the evening of August 10th, Cotter hung the barometer just in time to read the last measurement before dark. He then walked all night back to the main camp.

In 1979, Bruce Dodge, Kirby Wilcox, Bill Helvestine and I climbed Mt. Goddard from Wanda Lake, a desolate sheet just below Muir Pass. The mountain is shingled with loose slate, making the ascent arduous, although not dangerous. The Survey was correct that the peak commands a vast vista in all directions. There were far more peaks in view than we could identify. I did not have a camera. In 1983, I visited the lake again to make this photograph. The vantage point is Wanda Lake, looking south. Cotter and Spratt approached from the west (the right).

aiming. On ascending it, however, it was found to he about eight miles due south of the Obelisk. It was found to be 10,950 feet high and commanded a fine view. The base is granite, and it is capped with lava; from this point it was noticed that some of the peaks of the Obelisk group are covered with the volcanic rock in the same way.

"Eighteen miles northeast of this is the lowest gap or pass over the Sierra [Mammoth Pass] which occurs between Carson's and Walker's Passes, a distance of about 250 miles. An approximation of its height was obtained by an observation of the barometer on the peak ascended near Camp 195 at a point which was ascertained by levelling to be at about the same altitude as the pass itself. The result of the calculation gave 9200 feet as the height of the summit of the pass, which is considerably lower than the Mono Pass. Cattle have been driven across to Owen's Valley over this route, the north fork of the San Joaquin being crossed at a point much farther up than where our party traversed it, and where the canyon is not nearly so deep."

Having decided to "push out," the Survey returned to Jackass Meadow, where they struck cattle trails and headed for Clark's Station. The route seems to have been south of Iron Mountain and then between it and the Fresno Grove (Nelder Grove) of Big Trees, thence along Big Creek to the South Fork of the Merced to Clark's.

SUNDAY, AUGUST 21

Came on a long ride 23 miles and camped at the head of the Chiquito Joaquin. In a few miles struck a cattle trail and at last the camp of the men where we got word of Orr, who ha[d] passed several days ago. [D] Hoffmann had grown so much worse that by night he had to be lifted from his horse and could not walk a step. It had been cloudy all day, and soon after dark a cold rain set in and it rained hard all night. There is no need of again describing the discomforts and miseries of sleeping on the ground in a cold, rainy night, when the rheumatism creeps into every nook and joint of one's frame. It was hard on poor Hoffmann. [L]

Monday, August 22

Some rain in the morning but soon ceased and we came on 16 miles. Hoffmann still worse. Along back all day. [D] We had to lift him on and off his horse and travel on the slowest walk. We had trails most of the day, however. [L] Crossed two cattle ranches. Crane Valley a most beautiful one. Camped at Lewis Ranch. We got some green corn and potatoes from the proprietor—grand treat indeed. [D]

Tuesday, August 23

Hoffmann very bad, cannot walk a step. Came on to Clark's Ranch—15-16 miles, rough trail. Hoffmann in great pain. Once rode through a hornet's nest. His horse backed and jumped. Terrible for him.

At Clark's found Mr. and Mrs. Ashburner, Olmsted and associates, and King. [D] You can imagine the joyous meeting. We were a hard-looking set—ragged, clothes patched with old flour bags, poor—I had lost over thirty pounds—horses poor. [L]

King had big pile of letters for me. Over twenty and spent the rest of the day between reading letters and talking with my friends but did not finish either. [D]

After Galen Clark struck out as a goldminer, he worked as a packer and camp-keeper in Mariposa until he first saw Yosemite Valley in 1855. Within a year, Clark filed a land claim on a meadow on the South Fork of the Merced, a long day's ride south of the Valley. He began ranching there and discovered nearby the Mariposa Grove, which the Indians called Wawona. The Mariposa Trail was then constructed along the South Fork and Clark began to operate a hostel for travelers to the Grove and to the Valley.

Just before the Survey's visit, on July 1, 1864, President Lincoln signed historic legislation ceding the Valley and the Grove to California for preservation as a park. Shortly after the visit, Clark was appointed by Governor Low to the Yosemite Commission along with Whitney, William Ashburner, Frederick Olmsted and others. Clark was named by the commission to be the first Guardian of Yosemite. He lived at Wawona and in the Valley until his death in 1910, and is buried in the Valley. Mt. Clark, the Clark Range, and Clark Point (above Vernal Falls) were named for him.

As noted, King had agreed to rejoin the party at Clark's. Although his report to Whitney states that he arrived on August 1st at 10 a.m., the journal of Harriet Errington (the Olmsteds' governess) places the arrival on the morning of the second. Until the 23rd, according to Miss Errington, King accompanied the ladies to the Mariposa Grove, escorted them to Ostrander's and Yosemite Valley, and "amused us with details of his hair breadth escapes."

Wednesday, August 24

A crowd left for Mariposa, our boys. Mr. and Mrs. Ashburner stopped over. In p.m., Ashburner and I visited the grove of Big Trees known as the Mariposa Grove, although actually in Fresno County. (We passed near the Fresno Grove [Nelder Grove] yesterday.) Trees fine, grand, as that species always is, some on hills, some in valleys. A pleasant ride six miles from Clark's (called five) and 2400 feet above it with a grand view of the surrounding landscape. [D]

Thursday, August 25

Mr. and Mrs. Ashburner wished to start and I went with them to the Wagon Road ten miles [away] down tremendous hills leading her horse and riding back of his. Fine trail, grand views and immense forests. At the road met King, who had rode up in a wagon and had my letters, a pile of them. Also a watermelon, which proved delicious. I have a surfeit of letters, over thirty, in these two days. The boys returned from Mariposa and we took Hoffmann to the house and put him in a bed. [D]

William Ashburner (1831–87) was a member of the Survey who, for health reasons, had been unable to make the expedition. Shortly after this date, he too was appointed by Governor Low to the Yosemite Commission. After leaving the Survey, Ashburner became a prominent citizen of California, practicing mining engineering in San Francisco and serving as a regent of the University of California and as a trustee of Stanford University. Mrs. Ashburner was the daughter of Justice Stephen J. Field, who had been influential in establishing the Survey.

Yosemite Valley & Tuolumne Backcountry

FRIDAY, AUGUST 26

Mr. Olmsted came last night and at three I started with him for Yosemite. [D] He is the manager of the Mariposa Estate. His family [is] in the valley, where they spend several weeks during the heat of the summer. [L]

We rode 13 miles with grand views as we ascended of the forests and foothills and the great plain and the Coast Ranges. After dark when we got to Ostrander's, whose hospitable wife insisted in our taking supper although we had lunches with us. Then we built an evening fire and slept near it. [D]

Frederick Law Olmsted was America's pioneer landscape architect, a journalist, and public citizen. In 1858, he had enhanced his acclaim by winning, together with Calvert Vaux, the competition to design Central Park in New York. During the Civil War he served on the United States Sanitary Commission. In 1863, he resigned to become the superintendent of the Mariposa Estate, a principality of seventy square miles in the Sierra foothills near Yosemite, which boasted several gold mines and the camp towns of Mariposa and Bear Valley, where Olmsted lived.

The fortuitous meeting of Olmsted and Brewer occurred during Olmsted's first summer in Yosemite. On July 14, the Olmsteds and the Ashburners left Bear Valley for Clark's Station (near Wawona). There they spent several days, visiting the nearby Mariposa Grove and meeting King. On August 13, Olmsted, with his family, rode on to Inspiration Point and experienced for the first time that extraordinary vista of the Valley. Descending, they camped on the banks of the Merced opposite Yosemite Falls. After exploration of the Valley, Olmsted left on a hasty business trip to Bear Valley, his family remaining in the cool mountain air. On the return trip he met the incoming Survey on August 23 at Clark's.

Shortly after President Lincoln signed legislation ceding the Valley and the Mariposa Grove to California, Olmsted was rumored to be the choice to head the Yosemite Commission. In September Governor Low made it official. Thus, Olmsted's excursion with Brewer was the new commissioner's second visit to the Valley and his first exploration of the Yosemite backcountry. After the trip, Brewer wrote that "Mr. Olmsted is a very genial companion and I enjoyed it."

SATURDAY, AUGUST 27

Took our breakfast at the hospitable cabin of Ostrander and started. Rode about 17 miles. Trail as usual through forests of fir and *p. contorta* in the wet places. At last to Inspiration Point, a name not at all pedantic. View by far the finest of all about the valley. Day very clear and all spread about—the grand Cathedral, the grander Tutucanula above the emerald bottom. Beyond the rugged peaks and barren slopes toward Mono. Then descend and the rather tedious ride up the valley to Olmsted's camp just opposite the Fall. Pleasant evening with grand light on the Half Dome. [D]

SUNDAY, AUGUST 28

Most lovely morning, the Falls very slight, barely saw its shower of spray. The shadow of the Sentinel very sharp on the sides of the Cathedral. Wrote letter. Bid goodbye to my escort and started down the valley. Was to have been off by three but did not, of course—were off about sundown. [D]

The great Yosemite is ever grand, but it was less beautiful than when I saw it before. The season has been so very dry that the grassy meadows at the bottom were brown and sere, the air hazy, the water low. Over the great Yosemite Fall but a mere rill trickled, so small that it was entirely dispersed by the wind long before it reached the bottom of its great leap of half a mile. [L]

We rode down the valley about five miles until long after dark. We rode along by Tutucanula after dark. It loomed up against the sky like a great storm cloud. We slept by the river below the rock by the fork of the trail up the hill. [D]

On the same occasion, Olmsted recorded his rapture. The Valley, he said, "is sublimely beautiful, much more beautiful than I had supposed. The valleys is as sweet and peaceful as the meadows of Avon, and the sides are in many parts lovely with foliage and color. There is little water in the cascades at this season but that is but a trifling circumstance. We have what is infinitely more valuable, a full moon and a soft hazy smoky atmosphere with rolling, towering, white fleecy clouds."

MONDAY, AUGUST 29

Came on about 26 miles to Lake Tenaya. Up the long hill the grand views from near Yosemite Creek. [Stephen] Cunningham knows all the localities. He is guide and help, but plates, pepper and frying pans were forgotten. The latter brought down by Bell [Olmsted's cook] early in the morning. Two Indians have been with us but left. Cunningham can talk their language fluently. Camped at pretty Lake Tenaya with the grand scenery all around. Night decidedly cold. [D]

Olmsted and Brewer were accompanied by Olmsted's eleven-year-old stepson, John Olmsted, and Stephen Cunningham, a Yosemite resident and guide. They rode west out of the Valley via the Coulterville Trail and turned north, and then, east, via the Mono Trail.

TUESDAY, AUGUST 30

Came on about 19 miles. Stopped a short time at the Soda Springs and little farther up lunched. Met three men with a deer and bought half of it. Then on and camped near summit of Mono Pass at about 10,200 feet. The scenery all day very grand but not equal to that which we have seen about the head of Kings River. Both today and

30 BALLOON ROCK & THE MIDDLE FORK OF THE SAN JOAQUIN

On August 15th, "we packed up and started and made a big day's march toward the [middle] fork." The Survey followed the South Fork of the San Joaquin north until it joined the Middle Fork near what Brewer described as a "gigantic balloon struggling to get up through the rock." This scene looks from the Mile High Overlook east into the Middle Fork. The Survey approached from the right, crossed the gorge near Balloon Rock, and proceeded along the distant ridge to the left.

31 THE HEADWATERS OF THE MIDDLE FORK

Once across the San Joaquin, Brewer observed "many dark-colored peaks, apparently volcanic" on the crest to the east. These were Mts. Banner, Ritter and the Minarets, which form the headwaters of the Middle Fork. It was not until 1866 that King and Gardiner, leading another field party for Whitney, tried to climb these peaks. The snow fields of Mt. Banner empty into Thousand Island Lake, the scene of the photograph, looking eastward. The gull is from nearby Mono Lake, the breeding place of California gulls.

yesterday great banks of cumulus clouds on the Sierra at times and at others light cirrus far above the highest peaks. The air is very clear and cold. [D]

WEDNESDAY, AUGUST 31

Started for the summit but took the next peak south of Mt. Dana, fearing Olmsted could not reach the other. This I managed to get his horse up so that he rode to the top where we lunched. He named the peak Mt. Gibbs (after O. W. Gibbs) [a Harvard chemist and associate of Olmsted on the Sanitary Commission]. View very fine but inferior to that from Mt. Dana. Strange enough, we saw a group of persons on that peak clear against the sky. Only part of Mono Lake visible, a hill shutting out its center. Day was very cold. Returned and went down the valley six to eight miles and camped. [D]

Olmsted thought the view was good enough, writing a few days later to his father: "The view to the Eastward was very fine, the slope on that side being very abrupt, the desert plain of Mono 5,000 feet below us commencing more than six miles away. In the midst of the desert there was a considerable lake and three or four cones and craters of volcanic ashes. The horizon was everywhere broken by mountain ridges, those on the North East being in Nevada Territory and those in the South East beyond the valley of Owen's river, the most distant being more than 150 miles away. A few miles to the North of us was Mt. Dana, a very symmetrical peak 14,000 feet high; to the South, a group of peaks the center and highest being Mt. Lyell, which is a little higher than Mt. Dana. The Geological Survey spent several days trying to get to the top of it last year without success. On its North side there was a snow bank six miles long, in parts of which we could see the red snow described by Arctic travellers. Many Arctic plants and insects as well as birds and animals are found in this snowy region. We saw several. . . . Growing on and among these stones even to the very top, we found some beautiful alpine flowers most of them I believe indistinguishable from those found in the Alps."

WEDNESDAY, SEPTEMBER 1

Very cold. Fourteen degrees in the morning after sunrise. Started back and at the Soda Springs met Mr. St. John's party. They all visited Mt. Dana yesterday and all the party reached the top. Among the rest were a child, a little girl six years old and a man 69 years old and lame. We then crossed the river and struck onto the Mariposa Trail, a very rough trail indeed. The view of Cathedral Peak from the west [near Cathedral Pass] is peculiarly fine indeed. Passed several pretty little lakes and finally camped and set fire to a great dense tree which burned very fine. Night very cold and clear. [D]

Olmsted was cold: "The chief inconvenience of the whole trip was the cold weather at night. Though we made great fires and had all the blankets we could lie under, the cold kept me awake more or less every night. Every morning I found the water in my canteen under my pillow frozen."

THURSDAY, SEPTEMBER 2

Came on to Little Yosemite Valley above the Nevada Falls where we lunched. Then Cunningham and Olmsted went around while John and I came down over the Nevada Falls and got to camp long after dark tired enough. Found all well. [D]

This marked the end of Brewer's summer exploration. Before returning to Clark's, he visited the region above Mirror Lake with Mrs. Olmsted and studied the Valley Indians' baitless angling for trout in the Merced. Hoffmann's health remained a chief concern, especially after Brewer returned to Clark's on the 6th and found him no better. Brewer summarized the concluding days in his letter to his brother of October 12:

I got back to Clark's Ranch September 6 and found Hoffmann no better. He was anxious to get where he could get medical advice. At last, September 10, we four [King, Gardiner, Cotter, and Brewer] started to carry him ourselves. We made a litter and put a bed on it and

32 YOSEMITE PORTAL, MTS. HOFFMANN AND CONNESS

In 1864 the trail from Clark's (Wawona) ran closer to the rim than the road does today. Although somewhat off the old trail, Crocker Point has one of the finest views of the Valley and the high country. Mt. Hoffmann is the snow-clad peak in the center. Mt. Conness, named for the senator who introduced the bill to establish the park, is slightly to the right. The Cathedral Range is on the far right. These three regions were toured by Brewer and Olmsted at the end of the summer, and on August 27 Brewer and Olmsted passed near this view from Crocker Point.

started. The trail was so narrow that only two could carry at once. The trail led over a hill six thousand feet high [Chowchilla Mountain] and he grew heavier mile by mile, but it was successful. Gardiner and I returned for our animals, while King and Dick went on, and we all met again at Mariposa. There we got a carriage to Stockton, a hundred miles distant, and thence by steamer to San Francisco. He is still sick and may never recover.

Gardiner and I spent a few days on the Mariposa Estate, looking at its mines and mills, but I will give no description. It is a tremendous estate. I rode to Stockton with Mr. Olmsted in his private carriage, carrying $28,000 in gold bullion—quite a load.

I have received unofficial notice of my election as professor at Yale, and shall be on the road in a week if I can. I am now working hard to get off early, but will not close my journal yet, for a long trip still lies before me. I have counted up my traveling in the state. It amounts to: horseback, 7,564 miles; on foot, 3,101 miles; public conveyance, 4,440 miles—total, 15,105 miles. Surely a long trail!

Waiting for Brewer in the Montgomery Street offices of the Survey in San Francisco was a telegram received August 10 from Professor Brush of Yale, simply stating: "We want you sometime in October. Don't wait for Silliman."

---

33 CATHEDRAL ROCKS AND MEADOW

Although Brewer wrote that the "great Yosemite is ever grand," he noted that the "season has been so very dry that the grassy meadows at the bottom were brown and sere, the air hazy and the water low." This scene, of lush spring meadows in the Valley, stands in sharp contrast.

34 TENAYA LAKE AND MT. CONNESS

On the evening of August 29, Brewer and Olmsted "camped at pretty Lake Tenaya with the grand scenery all around." The first recorded exploration of Lake Tenaya had only been thirteen years earlier, in 1851, when the Mariposa Battalion went there to capture Chief Tenaya's band. Shortly thereafter, miners streamed past the lake to the mining districts near the Nevada border. Brewer's first visit to the lake was on June 25, 1863, with Hoffmann and Whitney. Later, King named the distant peak in this photograph for Senator John Conness of California.

35 MONO PASS AND BLOODY CANYON

By 1864 Mono Pass and Bloody Canyon were heavily traveled by miners on their way to the Esmeralda District. Brewer and Olmsted camped at the head of the pass on August 30. Brewer first visited Bloody Canyon in 1863 and wrote that it received its name from the fact that horses were so cut by sharp rocks that the rocks were "sprinkled with blood from the animals."

This is the scene from the head of the pass into Bloody Canyon toward Mono Lake below. For years, the water from this canyon was diverted to Los Angeles, along with the majority of water on the eastern slope. As a result, the water level in Mono Lake dropped steadily, so much that Negit Island, in the middle of the lake, was connected at times to the shore by a land bridge, thereby permitting predators to cross and destroy the eggs of California gulls. Litigation to stop the diversion finally succeeded after more than a decade, and the lake level has risen.

Postnote

Clarence King returned to Yosemite a few weeks after the summer exploration and led the Yosemite Boundary Survey (which was needed to define the borders of the federal grant establishing the park). In November 1864 he and Gardiner went east. They soon returned to the Survey. King then conceived of and obtained congressional approval of the Geological Survey of the Fortieth Parallel, which he led for several years beginning in 1867. In 1879, King was made the first director of the United States Geological Survey, a post he held until 1881. He subsequently practiced as a mining engineer in the West, Mexico, and Cuba; ranched; and wrote. His best known works were *Mountaineering in the Sierra Nevada*, *The Helmet of Mambrino*, and *The Age of the Earth*. When King died in 1901 he had become known throughout the land. His close friend, Henry Adams, said that, to his countrymen, King was "the ideal American they all wanted to be."

James Gardiner, born in 1842 in Troy, New York, was a boyhood and lifelong friend of Clarence King. After the 1864 expedition, he went on to assist King in the Yosemite Boundary Survey; accompanied him in the first ascent of Mt. Clark in 1866; and joined King in the Geological Survey of the Fortieth Parallel. After other survey work, Gardiner became the director of the State Survey of New York until 1886 and then practiced as a civil engineer in New York. As the director of the State Survey, he collaborated with Olmsted in the preservation of Niagara Falls. Gardiner's daughter wrote Francis Farquhar that, although King did not deserve such adoration, "my father worshipped him and served him devotedly all his life, as did John Hay, Adams, and his other close friends." Gardiner died in 1912.

Richard Cotter likewise assisted King in the Yosemite

Boundary Survey; went to Alaska; and then settled in Montana, where he became a miner, rancher, carpenter, justice of the peace and jack-of-all-trades. Cotter maintained a lifelong correspondence with Brewer, his letters recalling pleasant times from the "memorable summer of '64." Writing Brewer in 1898 that he had been appointed Sabbath School Superintendent, he said, "I imagine I can see you laughing as I have often seen in Camp."

Charles Hoffmann stayed with the Survey until its conclusion in 1874, briefly taught topographical engineering at Harvard, spent seven years in the Comstock Mine as the engineer of the Sutro Tunnel, mined with King in Mexico, and eventually settled in Oakland, where he developed an extensive and world-wide practice in mining engineering. The map-making technique pioneered by Hoffmann in California set the standard for decades to come. (That method consisted of establishing a series of precise triangles, duplicating the relative arrangement of the points on paper, and sketching in the terrain in between by eye.) Hoffmann also maintained a lifelong correspondence with Brewer. Hoffmann died in 1913.

William Brewer left California in November 1864 in response to his appointment to the Chair of Agriculture in the Sheffield Scientific School at Yale. He did not actually assume teaching responsibilities until the fall of 1865. In the meantime, he remained in Cambridge, Massachusetts, where he worked on the 1865 Report with Professor Whitney and began the cataloging of his botanical specimens from California. In his later career, he helped found the Connecticut Board of Health and served on it for thirty-one years. Brewer was a member of and elected president of the National Academy of Sciences and was the recipient of a number of honorary degrees. Brewer's enthusiasm for nature and conservation was lifelong. He joined in a student summer trip to the Rocky Mountains sponsored by Professor Whitney at Harvard; was part of an expedition to Greenland; and was a member of the 1899 Harriman Alaska Expedition. Brewer was appointed to the United States Forestry Commission. In 1896, with the assistance of John Muir, Brewer decried forestry waste and destruction in the lumber industry. Eventually, as a result of recommendations by the United States Forestry Commission and Brewer, the National Department of Forestry (now the Forest Service) was established with Gifford Pinchot as Chief Forester. Brewer married Georgiana Robinson in 1868. They had four children. Brewer died in 1910.

List of Camps

This list of the 1864 camps of the California Geological Survey, in the format below, is taken from a handwritten list contained in the Survey archives in the Bancroft Library. It is believed to have been made by William Brewer.

162.	Head of Cottonwood Creek (N. of Visalia)	June 8–9
163.	At Old Mill	June 9–10
164.	Thomas' Mill	June 10–17
165.	Eight miles ESE from Thomas' Mill	June 17–18
166.	Two miles below Big Meadows	June 18–24
167.	Two miles east of Granite Dome	June 24–25
168.	Little Meadow—By Divide	June 25–27 July 9–11
169.	South SW of Kettle	June 27–30
170.	By Granite Needle on Kettle Brook [Sugarloaf]	June 30–July 1
171.	Five miles west of Mt. Brewer	July 1–6
172.		
173.	King's Camps—Towards Mt. Tyndall	July 4–8
174.		
175.	South Fork of King's River [Roaring River]	July 6–9
176.	Big Meadows	July 11–17
177.	Seven miles east of Camp 176	July 17–18
178.	South Fork of King's River Canyon	July 18–19
179.	South Fork of King's River Canyon—At Head	July 19–21 July 25–26
180.	On Ridge North	July 21–25
181.	Near head of King's River	July 26–27
182.	Head of Little Pine Creek	July 27–28
183.	Near Bend City	July 28–29
184.	Black's Rock	July 29–30
185.	Warm Springs	July 30–31
186.	Round Valley	July 31–August 1
187.	West Fork of Owens River	August 1–2
188.	Head of San Joaquin River [Mono Creek]	August 2–4
189.	Middle Fork—18 miles below Camp 188 [Vermilion Valley]	August 4–6 August 13–15
190.	Near South Fork	August 6–9 August 12–13
191.	Near Mt. Goddard	August 9–12
192.	*Nearer* Mt. Goddard—Cold Camp!!	August 10–11
193.	Near North Fork of San Joaquin River	August 15–16
194.	On North Fork of San Joaquin River	August 16–17
195.	Flat 10 miles west of Camp 194	August 17–19
196.	Near Peak	August 19–21
197.	Head of Chiquito Joaquin	August 21–22
198.	Head of Fresno River	August 22–23
199.	Clark's Ranch	August 23–September 12
200.	Ostrander's	?? ??
201.	King's Camps	?? ??

36 VERNAL FALLS

Brewer and Olmsted crossed the Cathedral Range and returned via Little Yosemite Valley. Brewer hiked down via Nevada and Vernal Falls to the Valley.

37 TENAYA CREEK

Back in the Valley, Brewer escorted Mrs. Olmsted to Mirror Lake along pretty Tenaya Creek. Millions have since taken the same walk. About halfway up, there is an island in the stream. I crossed over to it one spring dawn. On the opposite bank of the far side of the island there was a beautiful stand of dogwoods.

38 WINTER STORM, YOSEMITE VALLEY

After helping carry Hoffmann by litter to Mariposa, King, Gardiner and Cotter returned to the high country around Yosemite, to survey the boundary of the new federal grant, until the first bitter storm of 1864 drove them out. King wrote Whitney: "About 10 a.m. for the first time during the autumn, a mild south wind came up and with it gray banks of snowy looking clouds. It is very hard to turn back when your object is near; I yielded to my constitutional sanguineness (when my judgment said return) and pressed on hoping for a favorable change of wind. . . . That night at eight o'clock the snow reached us and began to fall quietly. We tied our poor mules in as sheltered a place as possible, among the trees, and went to bed." The next morning, with more snow falling and poor visibility, the team plowed through the drifts back to the Valley, saving their lives, just barely. This scene and the next capture the aftermath of a winter snowstorm in the Valley.

39 NEW MORNING AND NEW SNOW, YOSEMITE VALLEY

This scene is from the vista point of Yosemite Valley made famous by Ansel Adams, the Wawona Tunnel Esplanade. Although this particular site was easy to reach by car, most of the scenes in this book required several days of hiking. Backcountry photographers might be interested in the following camera data. I have used two basic sets of gear, both utilizing field view cameras. Most of the scenes in this book were made with my lightweight backcountry set. This consists of a 4x5 Tachihara camera, which folds into a compact 2½-pound box, a Tiltall tripod, two lightweight lenses (a 300 mm Nikon lens and a 127 mm Ektar lens), 4x5 filmpacks, a focusing loupe, a featherweight level, assorted filters and accessories.

All but the tripod fit with room to spare into a daypack, which slips in turn into the upper compartment of my JanSport pack. The camera and lenses are wrapped in shock absorbent materials. My parka serves as a focusing cloth. To avoid the extra weight of a light meter, I have memorized the settings for Tri-X film for a series of recurring light conditions. On shorter backcountry trips, an additional lens is often added.

Some of the scenes in this book were taken with my heavier field set, suitable only for short day hikes. This includes a 5x7 Deardorff, a Zone VI "lightweight" tripod, three Schneider lenses (120 mm, 180 mm and 210 mm) and the 300 mm Nikon lens. This, together with additional accessories and a focusing cloth, print the total weight to about forty-five pounds.

List of Plates

The photographs appear in the approximate order that the scenes were encountered by the Survey in 1864. The frontispiece is listed in its proper chronological sequence. Dates given relate photographs to the narrative or editor's notes.

Sources

Brewer, William. Field Notes for 1864. (Bancroft Library).

This is a leather-bound pocket-size notebook used to record observations. It contains numerous barometric readings and some phrases and observations. Although it is not organized by date, some entries are dated.

Brewer, William. Letters to Edgar Brewer, 1860–1864. (Yale University Library).

Typescript copies of the letters are available at the Bancroft Library. These were published in *Up and Down California*, edited by Francis Farquhar, in 1930. In some instances, the letter text quoted here differs from the Farquhar version due either to rare errors in the original publication or to slight differences in editing.

Brewer, William. Pocket Diary for 1864. (Bancroft Library).

This is one of a series of leather-bound small pocket diaries, approximately three by five inches, with entries in pencil.

Brewer, William. Letters to G. J. Brush, dated January 22, 1864, and dated June 22, 26 and July 9, 1864. (Yale University Library).

These letters appear in the George J. Brush Papers at Yale. The latter was written by Brewer to Professor Brush at Yale in three installments during the 1864 expedition.

Brewer, William. Selected Correspondence. (Bancroft Library).

Brewster, Edwin Tenney. *Life and Letters of Josiah Dwight Whitney*. Boston: Houghton Mifflin Co., 1909.

California Geological Survey. *Geology, Vol. I, Report of Progress and Synopsis of the Field Work from 1860 to 1864.* Philadelphia: Caxton Press of Sherman & Co., 1865.

Chittenden, Russell. "Biographical Memoir of William Henry Brewer 1828–1910." In *Biographical Memoirs, Vol. 12.* Ed. National Academy of Sciences. Washington, D.C.: The Academy, 1929, pp. 135–169.

Errington, Harriet. "Harriet Errington's Letters and Journal from California 1864–65." Typed manuscript. (Yosemite Research Library).

Farquhar, Francis P. Correspondence and Research Materials for *Up and Down California.* (Bancroft Library).

Farquhar, Francis P. *History of the Sierra Nevada.* Berkeley: University of California Press, 1965.

Farquhar, Francis P. *Place Names of the High Sierra.* San Francisco: Sierra Club, 1926.

Gardiner, James. Selected Correspondence. (James Terry Gardiner Collection, New York State Library).

Hoffmann, Charles. Letter to J. D. Whitney. January 26, 1864. (Bancroft Library).

Keough, Thomas. "Over Kearsarge Pass in 1864." *Sierra Club Bulletin,* X (1918), 340–342.

King, Clarence. Field Notebook for 1864. (Huntington Library).

This is not organized as a diary, but King summarized segments of the 1864 trip, apparently within a few days of the events described. The segments include the trip to Thomas' Mill, the reconnaissance from there, the journey to Big Meadows, the ascent of Mt. Silliman, the ascent of "Mt. Whitney" (Mt. Tyndall), and the attempt on Mt. Whitney. This notebook contains King's "bearings from Mt. Whitney" [Mt. Tyndall] showing that Mt. Whitney was originally named Mt. Grant.

King, Clarence. "From Camp 118 to the Summit of Mt. Tyndall." 1865 Report to Professor Whitney. (Huntington Library).

This is a handwritten report on 8½-by 12-inch paper by King to Professor Whitney, used by the latter to complete the 1865 report entitled *Geology.*

King, Clarence. *Mountaineering in the Sierra Nevada.* Boston: James R. Osgood and Company, 1872.

Although the stories in this book are embroidered, the time and places indicated for the expedition itself conform to information available from other sources.

Olmsted, Frederick Law. Letter to John Olmsted. September 14, 1864. (Frederick Law Olmsted Papers, Manuscript Division, The Library of Congress).

Palmquist, Peter E. *Carleton E. Watkins, Photographer of the American West.* Albuquerque: University of New Mexico Press, 1983.

Sargent, Shirley. *Galen Clark, Yosemite Guardian.* San Francisco: Sierra Club, 1964.

U.S. Congress, Senate. *Reports of Exploration and Surveys to Ascertain the Most Practicable and Economical Route for a Railroad from the Mississippi River to the Pacific Ocean.* 33rd Cong., 2nd sess., Ex. Doc. No. 78. Washington, D.C.: GPO, 1856.

Wilkins, Thurman. *Clarence King; A Biography.* New York: Macmillan, 1958.

Acknowledgments

I WOULD LIKE to acknowledge the encouragement and many fine suggestions supplied by John Palmer, former Chief Park Interpreter of Sequoia and Kings Canyon National Parks; Jim Snyder, Yosemite National Park Service; and Henry Berrey and Steve Medley of the Yosemite Association. Mary Vocelka of the Yosemite Research Library found the Errington Diary and provided other leads. Barbara Lekisch of the Sierra Club Library led me to the Hoffmann Map and other Survey materials. Susan Acker of the Feathered Serpent Press was quite thoughtful in her original design of the volume. Susan and Wolf Schaefer of Phelps-Schaefer went beyond the call of duty in producing faithful lithographs. Finally, my wife, Suzan; my daughter, Allison; our good friends Kirby Wilcox, Bruce Dodge, Bill Helvestine, Joe Turnage, Brad Bishop and Tina Delli Gatti, whose companionship and good humor brightened numerous Sierra outings, made the search for the trail of the Survey a lasting memory.

WHA

Index

(c) denotes that the reference appears in a photo caption

SUCH A LANDSCAPE!

Composition and cover design for this edition of the book was by Sandy Bell of Springdale, Utah, using a Macintosh computer, based on original design by Susan Acker of the Feathered Serpent Press. The text was typeset in Bembo, with Chisel and Industrial 736 as display fonts. Printing by Phelps-Schaefer Litho-Graphics, Inc., Brisbane, California. Binding by Lincoln & Allen Bindery, Portland, Oregon.